T0339628

INTRODUCTION TO NATURE-INSPIRED OPTIMIZATION

INTRODUCTION TO NATURE-INSPIRED OPTIMIZATION

GEORGE LINDFIELD AND JOHN PENNY

ACADEMIC PRESS

An imprint of Elsevier

Academic Press is an imprint of Elsevier
125 London Wall, London EC2Y 5AS, United Kingdom
525 B Street, Suite 1800, San Diego, CA 92101-4495, United States
50 Hampshire Street, 5th Floor, Cambridge, MA 02139, United States
The Boulevard, Langford Lane, Kidlington, Oxford OX5 1GB, United Kingdom

Notices

Knowledge and best practice in this field are constantly changing. As new research and experience broaden our understanding, changes in research methods, professional practices, or medical treatment may become necessary.

Practitioners and researchers must always rely on their own experience and knowledge in evaluating and using any information, methods, compounds, or experiments described herein. In using such information or methods they should be mindful of their own safety and the safety of others, including parties for whom they have a professional responsibility.

To the fullest extent of the law, neither the Publisher nor the authors, contributors, or editors, assume any liability for any injury and/or damage to persons or property as a matter of products liability, negligence or otherwise, or from any use or operation of any methods, products, instructions, or ideas contained in the material herein.

Library of Congress Cataloging-in-Publication Data
A catalog record for this book is available from the Library of Congress

British Library Cataloguing-in-Publication Data
A catalogue record for this book is available from the British Library

ISBN: 978-0-12-803636-5

For information on all Academic Press publications
visit our website at https://www.elsevier.com/books-and-journals

Working together
to grow libraries in
developing countries

www.elsevier.com • www.bookaid.org

Publishing Director: Joe Hayton
Editorial Project Manager: Kattie Washington
Production Project Manager: Mohana Natarajan
Designer: Matthew Limbert

Typeset by VTeX

This book is for my wife Zena. With tolerance and patience, she has supported and encouraged me for many years.

George Lindfield

This book is for my grandchildren, Jake, Ted, Lois, Scarlett, Ben, Josh, Maia, Kate, Sadie and Connie, hoping that at least one of them may one day find this text interesting.

John Penny

CONTENTS

ABOUT THE AUTHORS

George Lindfield is a former lecturer in the Department of Computer Science at Aston University and is now retired. He taught courses in computer science and in optimization at bachelor- and master's-level. He has coauthored books on numerical methods and published many papers in various fields including optimization. He is a member of the Institute of Mathematics, a Chartered Mathematician and a Fellow of the Royal Astronomical Society.

John Penny is an Emeritus Professor in the School of Engineering and Applied Science at Aston University, Birmingham. England. He is a former head of the Mechanical Engineering Department. He taught bachelor- and master's-level students in structural and rotor dynamics and related topics such as numerical analysis, instrumentation and digital signal processing. His research interests were in topics in dynamics such as damage detection in static and rotating structures. He has published over 40 peer reviewed papers. He is a Fellow of the Institute of Mathematics and its Applications and is a coauthor of four books with Elsevier, Cambridge University Press, Prentice Hall and Ellis Horwood.

PREFACE

This introductory text on nature inspired optimization is suitable for undergraduate and postgraduate students, and we hope it will be of value to those working in various fields of science, engineering and finance who may be confronted with nonlinear optimization problems, where these methods may used to advantage. However, it is not written with researchers in field of optimization in mind.

As the title suggests, this book aims to introduce the reader to several of the innovative mathematical methods for non-linear optimization that are inspired by the natural world. Some of the methods are based on biological behavior and the way various species behave in order optimize their chances of survival. Other methods are inspired by the laws of physics.

We begin with discussions of optimization problems, both linear and non-linear, and introduce some classical optimization methods for their solution. These methods have limitations and we illustrate these difficulties. Most nature inspired algorithms are based on stochastic processes and we introduce some important random distributions.

The bulk of text (Chapters 2 to 8) is devoted to describing a range of specific methods and provides numerical studies based on accepted and widely used test functions. We give many graphical illustrations. The algorithms described include evolutionary algorithms such as the genetic algorithm and differential evolution; algorithms based on swarming such as the PSO, and a range of algorithms based on the behavior of fireflies, cuckoos, ants, bees and bacteria. We also describe a range of algorithms based on the laws of physics, such as the annealing process, the big bang expansion of the universe and the laws of gravity. An interesting feature of one physically inspired method we consider is that it is deterministic. At the end of these chapters we have provided problems for the reader to try. However, it is not easy to devise simple problems in this field because the methods of solution are both random and, for hand calculation, tedious.

In Chapter 9 we briefly introduce important classes of non-linear optimization problems. Specifically, integer programming in which solutions must be integers, constrained non-linear optimization and multi-objective optimization. In the closing chapter we briefly describe a small selection of promising algorithms. In this chapter we include limited comparative studies on four selected algorithms described in the text. We also provide MATLAB scripts for three algorithms to allow the reader to experiment with these algorithms and assess their performance on certain problems.

George Lindfield

John Penny

ACKNOWLEDGMENT

We thank Kattie Washington for her encouragement and support, Joe Hayton, the Publishing Director, Mohana Natarajan, the Production Manager and Matthew Lambert, Designer.

NOTATION

In this text we have endeavored, where every possible, to use a standardized notation, rather than describing algorithms in the notation used by the original authors. We use the following notation

n_{var}	Number of variable in the problem. The number of dimensions is identical.
n_{pop}	Size of the population of trial solutions, sometimes called agents or probes.
n_{gen} or t_{max}	Maximum number of generations or iterations, or epochs. When stated, superscript k gives the number of the current iteration or generation.
$x_{ij}^{(k)}$	Location of the jth trial solution in the ith dimension, at generation k.
$\mathbf{x}_j^{(k)}$	Location of the jth trial solution, at generation k.
$\mathbf{X}^{(k)}$	Location of all the trial solutions at generation k.
$f(\mathbf{x}_j^{(k)})$	Value of function f for the jth trial solution, at generation k.
$f_j^{(k)}$	Equivalent to $f(\mathbf{x}_j^{(k)})$.
$r_u, \mathbf{r}_u, \mathbf{R}_u$	A scalar, vector or array of random numbers taken from a uniform distribution of random numbers in the range 0–1.
$r_n, \mathbf{r}_n, \mathbf{R}_n$	A scalar, vector or array of random numbers taken from a normal or Gaussian distribution of random numbers with a zero mean and a unit standard deviation.
$r_l, \mathbf{r}_l, \mathbf{R}_l$	A scalar, vector or array of random numbers taken from a Lévy distribution of random numbers.

CHAPTER 1

An Introduction to Optimization

1.1 INTRODUCTION

Optimization is the task of finding the best solutions to particular problems. These best solutions are found by adjusting the parameters of the problem to give either a maximum or a minimum value for the solution. For example, in a mathematical model of a manufacturing process we might wish to adjust the process parameters in order to maximize profit. In contrast, in the design of a steel structure required to carry a particular load, we might adjust the design parameters to minimize the weight of steel used, thereby minimizing the material cost. In both examples we are seeking to find an optimum solution. The function for which the optimum value is being sought is called the objective function, fitness function or cost function. The latter name arises because the purpose of optimization is often to reduce costs. Note that optimizing a function means finding the values of the function parameters to give the optimal or best value of the function.

In the following sections we discuss some of the classes of optimization problems and methods of solution.

1.2 CLASSES OF OPTIMIZATION PROBLEMS

Optimization problems can usefully be divided into two broad classes, linear and non-linear optimization. We begin by discussing linear optimization. As the name implies, both the objective function and the constraints are linear functions. Linear optimization problems are also referred to as linear programming problems. (Here programming does not refer to computer programming.) The general form of this problem is give by (1.1) thus

$$\begin{aligned} \text{minimize } f &= \mathbf{c}^\top \mathbf{x} \\ \text{subject to } \mathbf{A}\mathbf{x} &= \mathbf{b} \\ \mathbf{x} &\geq 0 \end{aligned} \tag{1.1}$$

where $\mathbf{x}$ is a column vector of the n parameters that we wish to determine in order to minimize f. The constants of the problem are given by an m component column vector $\mathbf{b}$, an $m \times n$ array $\mathbf{A}$ and an n component column vector $\mathbf{c}$. Note that each element of $\mathbf{x}$ is constrained to be greater than or equal to zero. This is a common requirement in practical problems which reflects linear programming's real world application. For

Introduction to Nature-Inspired Optimization
DOI: 10.1016/B978-0-12-803636-5.00001-3
1

example, if a plant could manufacture 1000 tonne of chemicals of three types, an optimum solution to minimize production costs might be to manufacture 120.3 tonne of chemical 1, 539.8 tonne of chemical 2 and 339.9 tonne of chemical 3. This is a realistic solution. Obviously a negative output is impossible so that a solution of 884.6, 766.5 and −651.5 tonne, although algebraically equal to 1000 tonne, is impossible, or infeasible as it does not satisfy $\mathbf{x} \geq 0$ constraint.

Note that these equations assume that the m constraints are of the form $\mathbf{Ax} = \mathbf{b}$. If one or more of the constraint equations are of the form $\mathbf{A}_j\mathbf{x} \leq \mathbf{b}_j$, where $\mathbf{A}_j$ is the jth row of $\mathbf{A}$, then these equations can be converted to the standard form by adding so called 'slack variables' to the equations. Similarly, if $\mathbf{A}_j\mathbf{x} \geq \mathbf{b}_j$ it can be converted to the standard form by subtracting slack variable, sometimes called surplus variables, from the equations. The slack variables are added to the vector $\mathbf{x}$, that is they become extra degrees of freedom, or variables, in the problem.

All linear optimization problems are constrained. If there are no constraints the problem becomes trivial. For example, suppose we wish to find the maximum value of the linear unconstrained function

$$f(x, y) = 2x + 3y$$

Obviously the maximum value of the function is given when x and y tend to infinity. Similarly the minimum value of the function is given when x and y tend to minus infinity.

If constraints are imposed, such as $x, y \geq 0$ and $4x + 8y \leq 40$ then by simple reasoning we can see that the maximum value of the function will be obtained when $4x + 8y$ is equal to 40, not less than 40. By trial and error we find that the maximum value of the function $f(x, y)$ is 20 and this is obtained when $x = 10$ and $y = 0$. Note that these values of x and y satisfy the constraints, as they must do. In this example the unknown variables are x and y. Normally we will use subscripted variables for the unknowns, i.e. x_1, x_2 etc., since in real linear optimization problems may have hundreds or even thousands of variables and an equally large number of constraints. Powerful algorithms have been developed to solve these problems, for example Karmarkar (1984). However, this is not a major area of application for nature inspired optimization methods.

We now consider non-linear optimization problems. Here the objective function is non-linear and the constraints, if they exist at all, may be linear or non-linear. The general statement of a non-linear optimization problem is

$$\text{minimize } f(\mathbf{x})$$
$$\text{subject to } g_i(\mathbf{x}) \leq 0, \quad i = 1, 2, \dots p \tag{1.2}$$
$$h_j(\mathbf{x}) = 0, \quad j = 1, 2, \dots m$$

where $\mathbf{x}$ is a vector of n elements. If any inequality constraints are of the form $g_i(\mathbf{x}) \geq 0$ then they may be converted to $g_i(\mathbf{x}) \leq 0$ by multiplying the constraint equations by −1.

If the problem has no constraints it is called an unconstrained optimization problem. Non-linear problems may have many local optimum solutions, which are optimum in a specific sub-region of the solution space. However, the optimum in the whole region for which the problem is defined is called the global optimum. We can draw an analogy with a mountainous region of land. There are many peaks and valleys in the region but only one highest peak and one lowest valley. This is an important issue in optimization, and we will return to it frequently.

In some optimization problems the parameters can take any real values to give an optimal solution, but in others parameter can take only positive integer values. For example, the optimal number of people involved in completing a task can only be a positive integer. This class of problem is important. It is very challenging and is often called a combinatorial problem. It includes, for example, job scheduling and the traveling salesman problem.

We now briefly describe some of the possible ways in which non-linear optimization problems can be solved.

1.3 USING CALCULUS TO OPTIMIZE A FUNCTION

It might be thought that calculus could be employed effectively to find minimum or maximum vales of a function. This is true, but the number of differential coefficients required, and the complexity of the resulting algebra, for anything other than simple problems, rules out this approach for most practical applications.

We illustrate the general approach by consider the simple function

$$f(x, y) = x + 2y + 32/(xy)$$

Differentiating this function with respect to x and y and gives

$$\partial f/\partial x = 1 - 32/(x^2 y)$$
$$\partial f/\partial y = 2 - 32/(xy^2)$$

For a maximum or minimum value the partial derivatives must equal zero. Thus

$$1 - 32/(x^2 y) = 0, \quad 2 - 32/(xy^2) = 0$$

Simplifying the above equations gives $x = 2y$ and substituting for x in the first equation gives $y = 2$. Thus $x = 4$. In the classical calculus approach we determine whether this solution is a maximum or a minimum of the function by considering the second derivative of the function. Thus, for the above equations, this gives

$$\partial^2 f/\partial x^2 = 64/(x^3 y), \quad \partial^2 f/\partial y^2 = 64/(xy^3), \quad \partial^2 f/\partial xy = 32/(x^2 y^2)$$

Using these differential coefficients we can determine the parameter D thus:

$$D = \frac{\partial^2 f}{\partial x^2} \times \frac{\partial^2 f}{\partial y^2} - \left(\frac{\partial^2 f}{\partial xy}\right)^2$$

Substituting the values of x and y that make the first derivatives zero, $(x = 4, y = 2)$ then if $D > 0$ the function is a minimum, if $D < 0$ the function is a maximum and if $D = 0$ the function is a saddle point. In this case

$$D = 0.5 \times 2 - 0.5^2 = 0.75$$

Thus this solution is a minimum.

Consider a second example, that of finding the minimum value of the function $f(x, y) = (x^4 - 16x^2 + 5x)/2 + (y^4 - 16y^2 + 5y)/2$. This function is a famous standard test function called the Styblinski-Tang function, see Appendix A for information on this function. Differentiating this function with respect to x and y and setting the derivative to zero gives

$$\partial f / \partial x = (4x^3 - 32x + 5)/2 = 0$$
$$\partial f / \partial y = (4y^3 - 32y + 5)/2 = 0$$

These cubic equations are uncoupled since they each contains only one of the variables and may be solved independently of each other by any appropriate numerical procedure. Solving the first cubic equation in x gives the following three solutions for x:

$$x_0 = 0.1567 \quad x_1 = 2.7468 \quad x_2 = -2.9035$$

Obviously, by symmetry, the second equation gives identical values for y, i.e. we can say that $y_0 = x_0$, $y_1 = x_1$ and $y_2 = x_2$. The second derivatives, $\partial^2 f / \partial x^2$ and $\partial^2 f / \partial y^2$ are both positive for $x = x_0$, $y = y_0$ and so this point must be a maximum of the function. Substituting for x_0, y_0 in f gives a value of 0.3912. When $x = x_1$ or $x = x_2$ and $y = y_1$ or $y = y_2$ the second derivatives of the function are negative, so these values, taken in x and y pairs must give minimum values of the function. Substituting for x and y values in the function gives

$x_1 = 2.7468,$	$y_1 = 2.7468,$	$f = -50.0589$
$x_2 = -2.9035,$	$y_1 = 2.7468,$	$f = -64.1956$
$x_1 = 2.7468,$	$y_2 = -2.9035,$	$f = -64.1956$
$x_2 = -2.9035,$	$y_2 = -2.9035,$	$f = -78.3323$

Clearly, $f = -78.3323$ is the global minimum, the other three minima are local minima. Other combinations of x and y, for example x_0 and y_1 are saddle points. In this problem

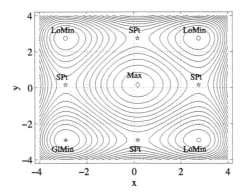

Figure 1.1 Contour plot of the Styblinski-Tang function showing a maximum, one global minimum, three local minimum and four saddle points of the function.

there are nine possible solutions, as shown in Figure 1.1 and each one has had to be considered as a possible global minimum, and eight rejected. In more complex problems this requirement could be onerous and the process not efficient.

Calculus methods can be extended to include the influence of constraints using, for example, the Lagrange multiplier method described in Chapter 9. However, the basic problem of the calculus approach remains that for any problem with more than a small number of variables, the algebraic manipulation and analysis becomes too complicated and inefficient.

1.4 A BRUTE FORCE METHOD!

A simple method of finding the optimal value of a function is to systematically evaluate the function for a large number of values. Suppose, for example, we require to determine the optima of a two variable function to an accuracy of one part in one hundred in both x and y search range. This requires 100×100 grid of function evaluations, $10,000$ in total. Using a modern computer, this takes only a fraction of a second to complete. An accuracy of one part in one thousand still only requires $1,000,000$ evaluations. However, determining an optimal solution to an accuracy of one part in one thousand in a problem with 10 variables would require $1000^{10} = 10^{30}$ evaluations. Table 1.1 provides some computing times required to evaluate Rastrigin's function (defined in Appendix A) in 2, 3 and 4 dimensions with a search range of -2.4 to 2.4. The computation times shown in Table 1.1 are obtained using a modest laptop computer running MATLAB. For Rastrigin's function in 4 dimensions or variables, where 1001 divisions of the search range is required, no data is provided because a rough estimate suggests that the time taken to complete the process will be in excess of 10 hours!

Table 1.1 Time taken (in seconds) to evaluate Rastrigin's function

Variables	101 divisions/variable	1001 divisions/variable
2	0.0098	0.7686
3	1.0204	576
4	85.995	No data

Suppose that a more powerful computer was 1000 times faster. This would be of little significance in the problem of 10 variables where 10^{30} evaluations are required! Clearly, this is not a realistic way to find the optimum value of functions in many variables.

Using 101 (or 1001) divisions in the search range provides the exact solution for the minimum of this function, which is $x_i = 0$, $i = 1, ..., n$ and $f_{min} = 0$. This is because one of the evaluation points is exactly coincident with this minimum. If, instead, the search ranges had been divided into 104 divisions then the closest we can get to the minimum in the 3 variable problem is $x_1 = x_2 = x_3 = 0.02330$. This gives an estimate of $f_{min} = 0.3226$.

This approach vividly illustrates the so called "curse of dimensionality". What seems an efficient method of solution in a problem in only two or three dimensions becomes extremely difficult if not impossible in a high dimensional problem.

1.5 GRADIENT METHODS

Section 1.3 of this chapter shows that a calculus approach can only solve small problems and multiple evaluations of the function to be optimized becomes impossible when the number of variables and possibly the required accuracy is increased. The methods described here, generally called gradient methods, have proved effective in finding optimal values of functions, but not necessarily global optimal values. If there are many local optima then the likelihood that the algorithm will find the global optima diminishes.

Although the function to be optimized may be a function of many tens or hundreds of variables, it is easier to visualize the problem if we consider a function of two variables. In this case, the objective function represents a surface as shown in Figure 1.1. Let us further assume that the problem is to find a minimum value of the function or surface. We assume that in this problem there are no constraints. If we imagine the objective function surface is a mountainous region and we were standing on high ground, how would we find the lowest level, the minimum value of the surface? Obviously walking in a direction which takes us down hill would be a good idea and this in essence is the basis of the steepest descent method used in minimization. This numerical procedure starts at some point on the surface of the function and moves to a lower value of the function in a series of steps until a minimum is found, i.e. further steps in any direction lead to higher values. Thus the method has found a minimum value of the function but it may not be the global minimum (i.e. the minimum of all the local minimum), it may merely be a local minimum. In the same way, if we walk down from high ground into a valley, we do not know whether or not we have found the lowest valley in the region, we have simply found a valley. This is a major difficulty with classical methods, such as steepest descent. Is the minimum value found a local minimum or the global minimum?

The steepest descent method is the simplest but least efficient classical numerical method. More efficient is the conjugate gradient method, Newton's method and various quasi-Newton methods. However, for all of these algorithms a major difficulty is that the global minimum of the function may not be found, only a local minimum. Nature inspired algorithms which are described later in this text seek to overcome this difficulty.

The problem to be solved is

$$\text{minimize } f(\mathbf{x}) \text{ for all } \mathbf{x} \in \mathrm{R}^n \tag{1.3}$$

where $f(\mathbf{x})$ is a non-linear function of $\mathbf{x}$ and $\mathbf{x}$ is an n component column vector. This, of course, is a restatement of the non-linear unconstrained optimization problem. These problems arise in many applications, for example in neural network problems where an important aim is to find weights in a network which minimize the overall difference between the output of the network and the required output.

The standard approach for solving this problem is to assume an initial approximation $\mathbf{x}^{(0)}$ and then to proceed to an improved approximation by using an iterative formula of the form

$$\mathbf{x}^{(k+1)} = \mathbf{x}^{(k)} + \lambda \mathbf{d}^{(k)} \text{ for } k = 0,\ 1,\ 2, \ldots \tag{1.4}$$

Clearly to use this formula we must determine values for the scalar λ and the vector $\mathbf{d}^{(k)}$. The vector $\mathbf{d}^{(k)}$ represents a direction of search and the scalar λ determines how far we should step in this direction. A vast literature has grown up which has examined the problem of choosing the best direction and the best step size to solve this problem efficiently. For example, see Adby and Dempster (1974). A simple choice for a direction of search to find a minimum is to take $\mathbf{d}^{(k)}$ as the negative gradient vector at the point $\mathbf{x}^{(k)}$. For a sufficiently small step value this can be shown to guarantee a reduction in the function value. This leads to an algorithm of the form

$$\mathbf{x}^{(k+1)} = \mathbf{x}^{(k)} - \lambda \nabla f\left(\mathbf{x}^{(k)}\right) \text{ for } k = 0,\ 1,\ 2,\ \ldots \tag{1.5}$$

where $\nabla f(\mathbf{x})$ is defined as $(\partial f / \partial x_1, \partial f / \partial x_2, \ldots, \partial f / \partial x_n)$ and λ is a small constant value. The minimum is reached when the gradient is zero, as in the ordinary calculus approach. This is called the steepest descent algorithm. We also assume that there exists only one local minimum which we wish to find in the range considered. The problem with this method is that although it reduces the function value, the step may be very small and therefore the algorithm is very slow. A formal method for choosing a step size which gives the maximum reduction in the function value in the current direction may be described as follows. For each k find the value of λ that minimizes

$$f(\mathbf{x}^{(k)} - \lambda \nabla f(\mathbf{x}^{(k)})) \tag{1.6}$$

This procedure is known as a line-search. The reader will note that this is also a minimization problem. However, since $\mathbf{x}^{(k)}$ is known, it is a *one-variable* minimization problem in the step size λ. Although it is a difficult problem, numerical procedures are available to solve it. Equations (1.5) and (1.6) provide a workable algorithm but it is still slow. One reason for this poor performance lies in our choice of direction $-\nabla f(\mathbf{x}^{(k)})$.

A significant improvement of the basic gradient search is the quasi-Newton or variable metric method. The method is used to improve the slow convergence rate of the ordinary steepest descent method by supplementing it with information supplied by the inverse of the Hessian matrix of second order derivatives. Thus the search direction used is:

$$\mathbf{p}^{(k)} = -\left(\mathbf{H}^{(k)}\right)^{-1} \mathbf{d}^{(k)}$$

where $\left(\mathbf{H}^{(k)}\right)^{-1}$ is current the inverse Hessian matrix at step k and $\mathbf{d}^{(k)}$ is the gradient vector. Although this does increase the convergence rate, the difficulty with this is it requires the user to supply all the second order derivatives and the program must calculate the derivatives at each stage and invert the matrix. A very costly computational process. Consequently the Davidon, Fletcher and Powell method was developed, initially by Davidon (1959) and modified by Fletcher and Powell (1963). The method provided an approximation for the inverse Hessian matrix which used only the first order partial derivatives and could be updated at each iteration. In the formula they provide the dividing terms are scalars and so it is a relatively low cost updating formula. It also has the crucial property that if the initial approximation to the inverse Hessian is positive definite then successive approximations are positive definite. This ensures that each step, if small enough, will produce a reduction in the objective function value.

We now consider an improvement on the basic steepest descent method, called the conjugate gradient method. A major weakness of the steepest descent method is that successive directions of search are orthogonal. This is a very inefficient direction of search. The conjugate gradient method takes a combination of the previous direction and the new direction to approach the optimum more directly. It uses the same step size choice procedure given by (1.6) so we must now consider how the direction vector is chosen in the conjugate gradient method. Let $\mathbf{g}^{(k+1)} = \nabla f(\mathbf{x}^{(k+1)})$ so that the basic formula for the conjugate gradient direction is

$$\mathbf{d}^{(k+1)} = -\mathbf{g}^{(k+1)} + \beta \mathbf{d}^{(k)} \tag{1.7}$$

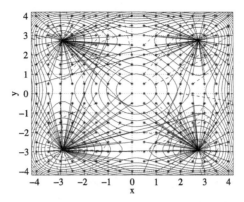

Figure 1.2 Steepest descent searches from a uniform grid of 17 by 17 starting values. Only start and end points, connected by straight lines, are shown. Global minimum is approximately $(-2.9, -2.9)$.

Thus the current direction of search is a combination of the current negative gradient plus a scalar β times the previous direction of search. The crucial question is: how is the value of β to be determined? The criterion used is that successive directions of search should be conjugate. This means that $(\mathbf{d}^{(k+1)})^{\top}\mathbf{A}\mathbf{d}^{(k)} = 0$ for some specified matrix $\mathbf{A}$.

This apparently obscure choice of requirement can be shown to lead to desirable convergence properties for the conjugate gradient method. In particular it has the property that the optimum of a positive definite quadratic function of n variables can be found in n or less steps. In the case of a quadratic, $\mathbf{A}$ is the matrix of coefficients of the squared and cross product terms. It can be shown that the requirement of conjugacy leads to a value for β given by (1.8):

$$\beta = \frac{\left(\mathbf{g}^{(k+1)}\right)^{\top}\mathbf{g}^{k+1}}{\left(\mathbf{g}^{k}\right)^{\top}\mathbf{g}^{k}} \tag{1.8}$$

Now (1.4), (1.6), (1.7) and (1.8) lead to the conjugate gradient algorithm given by Fletcher and Reeves (1964).

This method is an accepted, established and efficient algorithm with many applications but it is only guaranteed to find local optima.

The difficulty that classical methods have in finding a global optimum is illustrated in Figure 1.2. This figure shows the result of a relatively large number of searches performed to determine the minimum of Styblinski-Tang function in two dimensions. Only the start and end point of searches are shown with a straight line joining them. It can be seen that, depending upon the starting point, most searches fail to find the global minimum which lies at approximately $(-2.9, -2.9)$. Of course, the actual path taken by the searches would be much more complicated than the straight line shown in the figure.

Thus, it is apparent that even the best direct search methods do not guarantee that the global optimum will be found. Hence, there is a need for another type of optimization algorithm and this has led to the idea of nature inspired algorithms.

1.6 NATURE INSPIRED OPTIMIZATION ALGORITHMS

Purely mathematical methods such as steepest descent, conjugate gradient methods have been used for the optimization of non-linear functions. However some demanding problems such as global non-linear optimization and combinatorial problems have not been efficiently solved using these techniques and thus researchers have turned to other innovative techniques. Animal and plant life instinctively develop strategies and patterns of behavior over the millenia to ensure their survival when resources are precarious and scarce. In addition, some physical processes can be useful in developing optimization methods. For example simulated annealing, as its name implies, is based on the annealing process in which the rate of cooling of a material is controlled to improve the size of crystalline structure and reduce internal defects. It's clearly logical that the abundance and success of such strategies merit consideration in the development of algorithms for optimization, classification and related problems. These areas have provided the stimulus for a wealth of new ideas, particularly for solving demanding problems.

Almost all nature inspired algorithms have a random or stochastic element in their computation processes. This has a significant influence on the way the results from these algorithms are interpreted.

Nature inspired algorithms may be categorized as those inspired by biology and those inspired by natural science. The biologically inspired algorithms can be further subdivided into those based on evolution and those based on swarm behavior. The category of evolutionary algorithms includes the Genetic Algorithm and Differential Evolution. The swarm category includes Particle Swarm, Ant Colony, Artificial Bee Colony, Bacterial Foraging, Cuckoo Search and the Firefly algorithm. Algorithms based on the physical sciences include Simulated Annealing, the Gravitational Search algorithm and the Big Bang Big Crunch algorithm. A comprehensive list of nature inspired optimization algorithm is given by Fister et al. (2013). This list shows that in the last 25 years a very large number of algorithms have been developed. It is also reasonable to ask if some methods are demonstrably better than others. To this there is no simple answer. The No Free Lunch theory was introduced by Wolpert and Macready (1997). In this paper they stated the overall performance of optimization algorithms on a wide enough range of test problems would be same. Thus the performance of an algorithm might be very good for small or undemanding problems but it's performance for large and difficult problems would tail off and vice versa. This implies for a true comparison of algorithms a very wide range of test problems offering different challenges must be selected. Wolpert and McCready emphasize that they are making no judgments about the performance of individual algorithms. The individual user of optimization algorithms still has the task of selecting the best algorithm for their needs.

The types of problem these methods are used to solve are, in general, those which are computationally demanding and difficult problems with multiple optima where other methods lack effectiveness and efficiency and are attracted to local optima from which they cannot escape.

In contrast to deterministic algorithms, most nature inspired algorithms involve some element of randomness. In this case the algorithms provide different final results and follow different paths to the results each time they are run, even though the parameters for each run are the same.

The term metaheuristic algorithms is often used to describe the type of nature inspired algorithms we will consider. Metaheuristic is difficult to define, but it provides a top level structure giving a general procedure for solving a range of difficult problems. Thus it is a generalized method of solution, and is often inspired by nature. These algorithms in general provide solutions for demanding problems that have an acceptable level of accuracy and are determined in a realistic computation time.

Generally, nature inspired algorithms require two stages: random exploration of a large solution space and then exploitation which should provide a rapid and good convergence to the global optimum. This pair of processes are repeated iteratively until a satisfactory convergence to the optimum solution is obtained.

1.7 RANDOMNESS IN NATURE INSPIRED ALGORITHMS

The random element in these algorithms is introduced by selecting certain values from a random distribution. The starting point for the generation of random numbers is a specific computational method which generates what are called pseudo random numbers which is based on pseudo random number generator for example of the form:

$$x_{r+1} = (ax_r + b) \mod m \quad r = 0, 1, 2, \ldots \tag{1.9}$$

This is called a linear congruential generator where a and b and m are large integer constants that must be carefully chosen. Starting from some assumed initial value x_0 this equation can be used iteratively to generate a sequence of pseudo uniform random numbers. The mod operator simply means that after dividing $ax_r + b$ by m we take the remainder and assign it to x_{r+1}. The value of m determines the maximum number of pseudo-random numbers that can be generated before repeating.

However, hidden in this simple statement is great complexity since the choice of a, b and m is extremely difficult and a good choice is essential in obtaining acceptable uniform random values. Based on the complexities of number theory, the discussion of this is beyond the aims of this book. Fortunately MATLAB and many other computer languages provide suitable functions for the direct generation of pseudo-random numbers.

There are many different types of statistical distributions but we will concentrate on ones that are most commonly used in nature inspired algorithms to achieve specific types of random outcome which allows a random search of the region in which the problem is defined. These are the uniform random distribution, the normal distribution and the Lévy distribution. We only give a brief description of these distributions.

The uniform random distribution allows the generation of random numbers uniformly distributed over a specific range so that all outcomes are equally likely to occur. We denote the number selected from a uniform random distribution by r_u or $r_u(0, 1)$, where $(0, 1)$ indicates the range of the distribution. Some authors prefer to use U to denote a uniform distribution.

All distributions have a probability density function (pdf) denoted by $P(x)$ and a cumulative distribution function (cdf) denoted by $F(x)$ associated with them which defines their behavior. On a specific interval $[r, s]$ in the case of a continuous uniform random distribution these are defined by:

$$P(x) = \begin{cases} 0 & \text{if } x > s \\ \frac{1}{s-r} & \text{if } r \le x \le s \\ 0 & \text{if } x < r \end{cases} \qquad (1.10)$$

and the cumulative distribution function $F(x)$

$$F(x) = \begin{cases} 1 & \text{if } x \ge s \\ \frac{x-r}{s-r} & \text{if } r \le x \le s \\ 0 & \text{if } x < r \end{cases} \qquad (1.11)$$

Figure 1.3 provides a graph of the function.

The facility to generate random numbers with a uniform distribution is available in most computer languages. In particular, in MATLAB the function `rand(m,n)` allows the generation of a matrix of m rows and n columns of uniformly distributed random numbers in the range $[0, 1]$. This function can easily be adapted to generate uniform random numbers, y, in any range $[r, s]$ using the simple equation:

$$y = (s - r)x + r \qquad (1.12)$$

where x is a uniformly generated random number in the range $[0, 1]$. A key point here is that we can generate random numbers from any distribution for which we have a defined continuous cumulative distribution function, $F(x)$, by generating a uniformly distributed random number and setting it equal to $F(x)$ hence:

$$r_u = F(x)$$

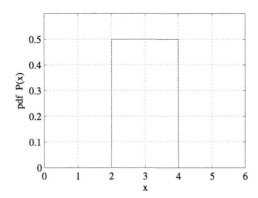

Figure 1.3 Uniform distribution. In this example the pdf is equal to 0.5 between 2 and 3 and zero elsewhere.

Consequently inverting we have:

$$x = F^{-1}(r_u)$$

as long as the inverse exists then x is distributed from the distribution defined by F. This is called the inversion method and can be used with any distribution for which we have an invertible cdf. There are many distributions and we now consider a few of the most relevant of these.

The normal distribution allows the generation of normally distributed random numbers and is one of the commonest distributions found naturally in the random processes of everyday life. In this text, we have denoted it by r_n or $r_n(\mu, \sigma)$ where μ, σ are the mean and standard deviation of the distribution respectively. Some authors prefer to denote a normal distribution by N.

The probability density function $P(x)$ for a normal distribution is defined as

$$P(x) = \frac{1}{\sigma\sqrt{2\pi}} e^{-\left(\frac{(x-\mu)^2}{2\sigma^2}\right)} \tag{1.13}$$

The graph of this function, taking $\mu = 0$ and $\sigma = 1$ called the standard normal distribution, is shown in Figure 1.4 and exhibits the shape often referred to as the bell curve.

Larger values of σ lead to flatter graphs and lower height varying values of μ shift the graph along the x-axis, positive values to the right and negative values to the left.

The cdf of the general probability function is given by:

$$F(x) = \left(\frac{1}{2}\right)\left[1 + \text{erf}\left(\frac{x - \mu}{\sigma\sqrt{2}}\right)\right] \tag{1.14}$$

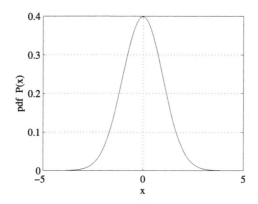

Figure 1.4 Normal probability density function with $\sigma = 1$ and $\mu = 0$.

Here the function erf is called the error function and is defined by the equation:

$$\text{erf}(x) = \frac{1}{\sqrt{\pi}} \int_{-x}^{x} e^{-t^2/2} dt \qquad (1.15)$$

Again using the inversion method we can generate random normally distributed values as we require.

The generation of normally distributed numbers can be achieved in MATLAB, for example, by using the function `randn(m,n)`. This function produces a matrix of m rows and n columns of normally distributed random numbers. The values generated follow the normal curve which is illustrated in Figure 1.4.

The final distribution we will discuss here is the Lévy distribution. Here we denote random numbers following this distribution by η or $\eta(c, \mu)$ where μ is the location parameter and c is the scale parameter. Other authors use a different symbol for the Lévy distribution. Some nature inspired algorithms use this particular distribution to obtain effective exploration of the region of search that hopefully ensures that global optimum is not overlooked. The definition of the Lévy probability density function is given by the function:

$$P(x) = \sqrt{\frac{c}{2\pi}} \left[\frac{\exp\left(-(c/(2(x-\mu)))\right)}{(x-\mu)^{3/2}} \right], \quad \text{for } x \geq \mu \qquad (1.16)$$

and the cumulative density function is defined by:

$$F(x) = \text{erfc}\left(\sqrt{\frac{c}{2(x-\mu)}} \right) \qquad (1.17)$$

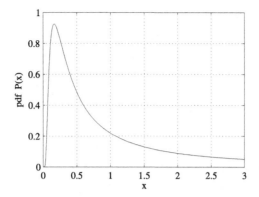

Figure 1.5 Lévy probability density function, with $c = 0.5$ and $\mu = 0$.

where erfc is a modified version of the error function called the complementary error function where $\mathrm{erfc}(x) = 1 - \mathrm{erf}(x)$. The graph of the Lévy distribution with $\mu = 0$ and $c = 0.5$ is given in Figure 1.5.

Note that unlike the normal distribution the Lévy distribution is clearly not symmetric and the right hand tail, rather than trailing off rapidly towards zero remains fairly large. Sometimes this is called a 'fatter' tail. The value of c lowers or raises the peak of the distribution producing a sharper or flatter distribution. Low values of c producing a sharp peak while large values produce a flatter curve. Using the inversion method values distributed according to the Lévy function can be generated.

These distributions allow different methods for exploring the region in which the problem is defined and to generate initial random populations. The heavy use of these random distributions will naturally lead to a variability in the outcomes of the runs of the nature inspired algorithms when used to solve specific problems and this important feature must be taken account of when testing the algorithms.

1.8 TESTING NATURE INSPIRED ALGORITHMS

In a rapidly developing field such as this it is important that all algorithms are tested thoroughly, equitably and intelligently. This a not a simple task for several reasons. In particular, because of the random nature of the processes involved, many test runs of the algorithms should be performed to smooth out the effects of this randomness and an average result obtained.

Many published tests of algorithms provide an average result and the standard deviation together with the best and worst result to clarify the nature of the performance of the algorithm. In comparing the performance of a number of algorithms, we face many problems. These are the efficiency of the computer code used, the random nature of the algorithms themselves and equitable setting of parameters at a fair and comparative level. This last difficulty needs some explanation. Suppose we wish to compare the firefly algorithm with simulated annealing. These two methods of optimization are quiet distinct and are described in Chapter 5 and 8 respectively. In the firefly algorithm we must set the number fireflies and the parameters α and β_0; in simulated annealing we must to set a quenching factor and an initial temperature. How do we choose these parameters to allow a fair comparison of the two methods. Obviously, a badly chosen quenching factor or initial temperature for the particular test problem would cause the simulated annealing algorithm to perform poorly and this would not provide an equitable comparison of the methods. It is not impossible to compare methods, but it does have to be done with care. A further issue is that some algorithms may perform very well on one problem, but not another, or may perform well on medium size problems but less well on large problems. Consequently a good set of standard test problems must be used which reflect the varying difficulties of nonlinear optimization.

In comparative studies tests of statistical significance, such as Friedman's test, may be useful. For details of Friedman's test see Daniel (1990). In addition the tests should be performed using the same parameters and measures of efficiency can be obtained using time measures or function evaluations. Function evaluations provide a better comparison between tests in different computing environments.

A key feature in testing nature inspired algorithms is the test problems used. These should test the key features of the algorithms. The dominant feature is that algorithms should be able to obtain the global optimum of a function. Thus problems which have many local optima should be part of the test set. The distribution of these optima is also a key feature. Some test problems should provide tightly packed optima others widely spread isolated optima since both types of function need to be optimized and they represent different challenges. The shape of the optima valley or peak should also be considered from a sharp almost discontinuous approach to the optima or a slow descent to poorly defined minimum. Fortunately a good and accepted set of test problems has been developed which reflects these features and others and new algorithms are often tested on this set, see Appendix A.

When running these tests it may come as a surprise to the reader that accurate results may not be found. Some of the problems are designed to be pathologically difficult and consequently only very approximate values may be obtained for the optimum. The dimension of the problem, i.e. the number of parameters in the problem, will influence the search. In these cases we are looking for the algorithm which gives the best result and some researchers in the field normalize the results to display this.

Since this book is not a series of research studies we have avoided tests on very large problems. Smaller problems allow us to illustrate the key features of the algorithm which can be replicated in reasonable time by the reader, even using a modest computer.

The MATLAB implementations provided in Appendix B are neither perfect or definitive or guaranteed for any purpose and are supplied only as a useful feature for readers. Furthermore, versions of algorithms can often be download from the internet, frequently provided by the original algorithm author. The tests we have run are not as exhaustive as those used in published research papers but provide an accessible useful flavor of how the tests may be performed by the reader.

1.9 SUMMARY

In this chapter the nature of optimization problems and the difficulties that arise when finding global optima are outlined. We describe the classic approach to solving such problems and their limitations. This naturally leads to the introduction of nature in-spired methods of optimization, the random nature of these methods and the relevant statistics. The chapter concludes with a discussion on the difficulties of fairly testing the performance of algorithms.

1.10 PROBLEMS

1.1 Derive an algebraic expression for the gradient of the following functions:

$$f_1 = x_1^2 + x_2^2 + x_3^2$$
$$f_2 = \cos(x_1^2 + x_2^2)$$
$$f_3 = 1/(x_1^2 + x_2^2 + 1)$$

Find the values of the gradients at $x_1 = x_2 = x_3 = 1$.

1.2 Taking $\mathbf{x}^{(0)} = [1, 2]^\top$ and using the function $f(x) = x_1^2 + 2x_2^2$ calculate an improved value for $\mathbf{x}$ from $\mathbf{x}^{(1)} = \mathbf{x}^{(0)} - \lambda \nabla f(\mathbf{x}^{(0)})$ in terms of λ. We determine λ from

$$\lambda = \min_\lambda(f(\mathbf{x}^{(1)}))$$

Show that this leads to the minimization of $f(\lambda) = (1 + 2\lambda)^2 + 2(2 + 8\lambda)^2$.

1.3 Find the minimum of $f(\lambda) = (1 + 2\lambda)^2 + 2(2 + 8\lambda)^2$ using calculus. Hence find the improved value $\mathbf{x}^{(1)}$ using $\mathbf{x}^{(1)} = \mathbf{x}^{(0)} - \lambda \nabla f(\mathbf{x}^{(0)})$. Compare the value of $f(\mathbf{x}^{(1)})$ with $f(\mathbf{x}^{(0)})$.

1.4 Perform two iterations of the gradient search method to find an approximation to the minimum of

$$f(\mathbf{x}) = (x_1 - 1)^2 + 3(x_2 - 2)^2$$

Take $\mathbf{x}^{(0)} = [0, 0]^\top$.

1.5 Use the conjugate gradient method to minimize the function $f(\mathbf{x})$ given in Problem 1.4.

1.6 Generate pseudo-random numbers using the expression

$$x_{r+1} = ax_r + b \quad \mathrm{mod}\ m$$

where $a = 7$, $b = 11$ together with both $m = 6$ and $m = 9$. Take $x_0 = 1$. What do you notice about the sequences of numbers you have produced.

1.7 (a) Determine an equation for the minimum value of the function

$$y = \sin(10\pi x)/(2x) + (x - 1)^4$$

(Gramacy and Lee, 2012) for $x \in [0.5,\ 2.5]$ by using calculus, together with a check on the function values at the boundaries. You are not expected to solve the equation you have derived.

(b) Use calculus to determine an equation for the minimum value of

$$y = (6x - 2)^2 \sin(12x - 4)$$

(Forrester et al., 2008) for $x \in [0,\ 1]$. You are not expected to solve the equation you have derived.

1.8 Use calculus to determine the minimum values of the function

$$z = x^3 + 3xy^2 - 30x - 18y$$

using the methods of Section 1.3 of this chapter.

1.9 The Dixon-Price function is a well known test function for global optimization methods. It is defined as

$$f(\mathbf{x}) = (x_1 - 1)^2 + 2(2x_2^2 - x_1)^2$$

and has solutions at $x_1 = 1$ and $x_2 = \pm 1/\sqrt{2}$ where $f(\mathbf{x}) = 0$. Using calculus verifies that this is the correct solution for this problem.

CHAPTER 2

Evolutionary Algorithms

2.1 INTRODUCTION

Evolution in nature tends to improve the overall fitness of the population through the generations and researchers noted that this concept could be applied to the optimization of nonlinear functions. Evolutionary algorithms refer to a group of optimization algorithms that are inspired by the process of evolution in human and animal life. The original algorithm is the Genetic Algorithm which takes the idea of chromosomes, comprised of genes, and models their life cycle. In the process of birth chromosomes are mixed, usually by crossover, and random mutation can occur. The successful chromosomes then give birth to a new, better generation whereas the poor chromosomes die. A more recent interpretation of the evolutionary process for optimization is the Differential Evolution algorithm. In this algorithm, evolution and hence optimization occurs through combining members of the population in a manner distinct from the Genetic Algorithm, together with a random mutation.

2.2 INTRODUCTION TO GENETIC ALGORITHMS

A specific example of evolutionary programming for optimization is the genetic algorithm. This algorithm is probably the oldest nature inspired method for optimization. It was initially developed by Holland and others in the 1960s and 1970s. The principal reference is Holland (1992). The basis of the method is to mimic in a simple form the genetic process of evolution. In all living things, genes hold the information to build an organism's cells and pass genetic traits to offspring. In the animal kingdom mating takes place and the offspring inherits some genes from the mother and some from the father. If the inherited genes make the offspring fit and strong it is more likely to survive and become a parent, passing on its good genes to the next generation. In contrast, an offspring with poor inherited genes is likely to be killed or die and not reproduce. This corresponds to the process of natural selection in nature. In this way, the population as a whole evolves and gets stronger; it becomes more successful and better suited to its environment and the species survives and flourishes. However, this process reduces the genetic diversity of the population and to compensate for this some genes mutate randomly, i.e. they change. A change may weaken a member of the population and in this case it is likely to die out. If the change makes the member of the population fitter, then it will survive and pass the modified gene on to future generations. In the animal kingdom, this process takes place slowly over many hundreds or thousands of years, but

Introduction to Nature-Inspired Optimization
DOI: 10.1016/B978-0-12-803636-5.00002-5

we can see the process working much faster in the way micro-organisms have developed a resistance to antibiotics. A natural population of bacteria contains a wide variation in its genetic material. When exposed to antibiotics, most bacteria die quickly, but some may have mutations that make them slightly less susceptible to the antibiotics. If the exposure to antibiotics is short, these individuals will survive the treatment. Due to the elimination of the vulnerable individuals in the past generation, this population will contain a higher proportion of bacteria that have some resistance to the antibiotic. At the same time, new mutations occur, contributing new genetic variation to the existing population. The populations of bacteria are so large that inevitably a few individuals will have beneficial mutations. If a new mutation reduces their susceptibility to an antibiotic then these individuals are more likely to survive when next confronted with that antibiotic. Over a period of time, a population of antibiotic-resistant bacteria will emerge.

The standard genetic algorithm (GA) for optimizing a function is based on a binary coded genetics, as follows. We randomly create a relatively small population of numbers, and initially we will assume the numbers are expressed in binary form. The numbers are frequently called chromosomes and the binary digits from which they are composed are referred to as genes. Thus, each member of the population comprises a single chromosome. Each of the chromosomes is a trial solution to the problem. We can convert each trial binary number to decimal and evaluate the function to be optimized. For each member of the initial population, the function value is a measure of its fitness. Members of the population are now chosen randomly to mate but the random choice is biased in favor of the fittest members of the population. The mating is usually performed by interchanging binary digits between a pair of chromosomes. This produces two new chromosomes (the children). These new chromosomes comprise a mixture of the parents' genes. After mating the parents die and are replaced by new generation of children and so in each generation, the size of the population remains constant. However, some of the children may be fitter than their parents and provide a better solution to the optimization problem. Finally some binary digits are randomly changed from 0 to 1 or vice-versa, usually with a low probability. This mimics the biological mutation process. In summary we have

BOX 2.1 Summary of GA
Step 1: Generate chromosomes from random numbers
Step 2: Determine fitness and select the fittest for mating
Step 3: Mate by crossover (recombination) to create offspring
Step 4: Introduce a low likelihood of mutation
Step 5: If the specified number of generations is not completed, go to step 2, else end.

We now look at each of these steps in detail. To illustrate how a basic GA can be used to solve an optimization problem, we now consider the task of minimizing the function

Table 2.1 Initial generation of binary digits with corresponding decimal vales and fitness values

Ref	binary	x_1	x_2	$f(\mathbf{x})$	$g(\mathbf{x})$	%
#0	100101|00000100	0.3492	−2.9059	8.5661	0.1045	2.1
#1	001001|11101100	−1.4286	2.5529	8.5583	0.1046	2.1
#2	100111|10110110	0.4762	1.2824	1.8712	0.3483	7.1
#3	100100|11000111	0.2857	1.6824	2.9119	0.2556	5.2
#4	011010|01101100	−0.3492	−0.4588	0.3325	0.7505	15.3
#5	101011|01100000	0.7302	−0.7412	1.0825	0.4802	9.8
#6	100001|01111010	0.0952	−0.1294	0.0258	0.9748	19.9
#7	100010|10111100	0.1587	1.4235	2.0516	0.3277	6.7
#8	100000|01111010	0.0317	−0.1294	0.0178	0.9826	20.1
#9	100101|10100010	0.3492	0.8118	0.7809	0.5615	11.5
Total				26.1986	4.8904	100

$f(x_1, x_2) = x_1^2 + x_2^2$. For this example, we take the search range to be $-2 \leq x_1 \leq 2$ and $-3 \leq x_2 \leq 3$. Obviously, the solution is when $x_1 = x_2 = 0$. We begin by generating a population of 10 chromosomes or trial solutions, each chromosome consisting of 14 binary digits, 6 to code the value of x_1 and 8 to code the value of x_2. The choice of 6 digits x_1 and 8 for x_2 is arbitrary, although it does control the maximum resolution in each variable. The binary digits 0 and 1 are generated with equal probability. For this example, we choose a mating rate of 0.6 (i.e. a 60% of the chromosomes are chosen to mate) and a mutation rate of 0.05 (i.e. a 5% chance of a gene mutating).

Each chromosome is a trial solution x_1 and x_2 for the optimum of the function $f(x_1, x_2)$. For x_1 we have chosen to use 6 binary digits and so the equivalent decimal values lie in the range zero to $2^6 - 1 = 63$. Multiplying by 4/63 scales these values to the range 0 to 4; subtracting 2 from each number scales the values to the range -2 to 2, the search range. Thus the smallest increment in x_1 is $4/63 = 0.06349$. We cannot get a more accurate estimate for the true value of x_1. For x_2 we have chosen to use 8 binary digits and hence the equivalent decimal values are in the range zero to $2^8 - 1 = 255$. Multiplying by 6/255 scales these values to the range 0 to 6; subtracting 3 from each number places them in the range -3 to 3. Thus the smallest increment in x_2 is $6/255 = 0.02353$, thus limiting the accuracy for x_2. Clearly, the choice of the number of binary digits used to represent x_1 and x_2 controls the accuracy of the solution. If we increased the length of the chromosome so that it comprised more binary digits we can obtain a more accurate estimate to the true values of x_1 and x_2. However, increasing the length of the chromosome increases the computation effort.

The 10 chromosomes we have generated are shown in the Table 2.1. The table also shows the scaled decimal value for each chromosome and the corresponding value of $f(\mathbf{x})$ where $\mathbf{x} = [x_1 \ x_2]$. If we were seeking the maximum value of the function then

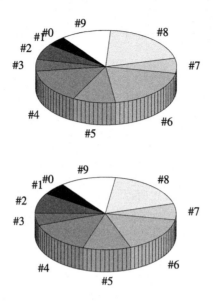

Figure 2.1 Basic roulette wheel. Top, $g = 1/(1 + f)$; Bottom, Proportional to Rank.

$f(\mathbf{x})$ would represent the fitness of each chromosome: the higher the value of $f(\mathbf{x})$, the fitter the chromosome. However, here we are seeking the minimum value of the function. To do this we define a second function $g(\mathbf{x}) = 1/(1 + |f(\mathbf{x})|)$. Clearly when $|f(\mathbf{x})|$ is a minimum $g(\mathbf{x})$ is a maximum and vice versa. Thus the fit chromosomes have high values for $g(\mathbf{x})$, the weaker chromosomes have lower values of $g(\mathbf{x})$. Note that the form of $g(\mathbf{x})$ is somewhat arbitrary. For example we could define $g(\mathbf{x}) = 13 - f(\mathbf{x})$. The constant 13 is chosen because it is the highest value of the function in the search region, i.e. $2^2 + 3^2 = 13$, so $g(\mathbf{x})$ would be in the range zero to 13.

We require the fittest chromosomes to mate in order to pass on their good genes to the next generation and in this process the chromosomes are considered to be genderless. However, we do not simply choose the fittest chromosomes for mating, but rather bias a random choice in favor of the fittest chromosomes. We see in Table 2.1 that the sum of the fitnesses, $g(\mathbf{x})$, is 4.8904. From this we can easily determine the percentage that each chromosome makes to the total fitness. It is these percentage contributions to the overall fitness, see Table 2.1, that are used to bias the random choice in favor of the fittest chromosomes. This is sometimes called the roulette wheel. The roulette wheel for this set of chromosomes is shown in Figure 2.1. Choosing the chromosomes to mate is analogous to spinning this biased roulette wheel. We observe that #4, #6 and #8 are the fittest chromosomes and they have a correspondingly greater chance of being selected. Note that if we had used the alternative definition of $g(\mathbf{x})$ given above, the percentage contribution each chromosome makes will be slightly different but the ranking will be the same.

Table 2.2 Biased random selection and mating six chromosomes

Ref	Selected	$g(x)$	Crossover	New	$g(x)$
#6	10000 ‖ 101111010	0.9748	after 5th	100001|01100000	0.6417
#5	10101 ‖ 101100000	0.4802	digit	101011|01111010	0.6452
#5	10101 ‖ 101100000	0.4802	after 5th	101011|00000100	0.1002
#0	10010 ‖ 100000100	0.1045	digit	100101|01100000	0.5983
#0	100 ‖ 10100000100	0.1045	after 3rd	100010|01101100	0.8093
#4	011 ‖ 01001101100	0.7505	digit	011101|00000100	0.1056
#6	10000101111010	0.9748		100001|01111010	0.9748
#6	10000101111010	0.9748		100001|01111010	0.9748
#5	10101101100000	0.4802		101011|01100000	0.4802
#0	10010100000100	0.1045		100101|00000100	0.1045
Total		5.4290			5.4347

Table 2.2 shows the chromosomes that have been randomly chosen for mating. On this occasion we see that #0, #5 and #6 have all been chosen three times, and #4 once but because of the random nature of the selection this will not always be the case. However, the total fitness of the selection is slightly higher. It is now 5.4290 compared to 4.8904.

We now carry out the mating process to generate the population for the next generation. We have set the mating rate as 0.6 and so 6 out of the 10 chromosomes are randomly chosen for mating. Mating is by a single point crossover. A crossover point is randomly chosen and can be after the 1st, 2nd, ... or 13th binary digit in our chromosome of 14 binary digits. The genes from the two parents are crossed over after the crossover point. Since 6 chromosomes have been chosen for mating, there must be 3 matings and in this example the crossover has been chosen randomly to occur at a point after the 5th, 5th and 3rd binary digit, as shown in Table 2.2. Note that new information has been produced since new values have been generated. If we consider the mating of chromosomes #0 and #4, crossing over after the 3rd binary digit, then expressed as decimal equivalents, we have

$$\text{Parents} \quad [+0.3492, -2.9059] \quad \text{and} \quad [-0.3492, -0.4588]$$
$$\text{Children} \quad [-0.1587, -2.9059] \quad \text{and} \quad [+0.1587, -0.4588]$$

Note that it is just coincidence that the two parents chosen for mating have identical values of x_1. Because crossover occurs in the genes defining the value of x_1, new values of x_1 are generated, namely 0.1587 or -0.1587. In contrast there is no crossover in the genes defining x_2 and so the values of x_2 are unchanged by this mating. After mating

Table 2.3 The population before and after mutation

Ref	Before mutation	$g(x)$	After mutation	$g(x)$
#0	100001|01100<u>000</u>	0.6417	100001|01100<u>110</u>	0.7304
#1	101011|01111010	0.6452	101011|01111010	0.6452
#2	101011|00000100	0.1002	101011|00000100	0.1002
#3	100101|01100000	0.5983	100101|01100000	0.5983
#4	100010|01101100	0.8093	100010|01101100	0.8093
#5	011101|0000<u>0</u>100	0.1056	011101|0000<u>0</u>000	0.0997
#6	100001|01111010	0.9748	100001|01111010	0.9748
#7	<u>1</u>00001|01111010	0.9748	<u>0</u>00001|01111010	0.2100
#8	101011|<u>0</u>1100000	0.4802	101011|<u>1</u>1100000	0.1495
#9	10010<u>1</u>|00000100	0.1045	10010<u>0</u>|00000100	0.1050
Total		5.4347		4.4224

the total fitness in this example has slightly risen to 5.4347. Sometimes it may decrease, but this is part of the overall exploratory nature of the process.

Finally we allow mutation to occur; i.e. each binary digit is examined and based on the selected probability is flipped from 0 to 1 or 1 to 0. In this case we have chosen a probability of mutation of 0.05. Since we have 10 chromosomes, each having 14 binary digits, there are 140 binary digits in the procedure. Thus we might expect about 7 of them to be changed. In fact, 6 binary digits have changed. The changed binary digits are shown underlined in Table 2.3. Note that after mutation the overall fitness of the population has fallen. This is not surprising since mutation is a random process and can raise or lower the overall population fitness. However the genetic diversity has increased, and each of the 10 chromosomes now have different values, enhancing the exploratory part of the process. This completes the first generation of the search and the second generation begins using the mutated population shown in Table 2.3. This process is continued over a large number of generations until the trial solutions have converged to an acceptable optimum.

To illustrate the operation of the binary GA we seek the optimum value, in the search range $0 \leq x \leq 1$, of the one dimensional function

$$F(x) = \exp\left[-2\log_e(2)\left(\frac{x - x_0 - 0.08}{0.854}\right)^2\right]\sin^6\left[5\pi\left\{(x - x_0)^{0.75} - 0.05\right\}\right]$$

Here we take $x_0 = 0$. This function, which is frequently used as a benchmark function to test optimization algorithms has, in the search range, five unevenly spaced maximum located at $x = 0.0796875$, 0.246, 0.45, 0.681 and 0.934. The value of the function at these values of x decreases from 1 (the global maximum) to 0.25. The binary GA is

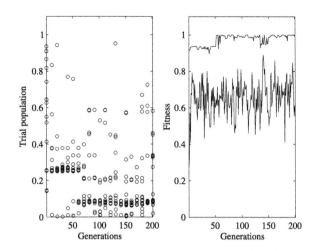

Figure 2.2 Optimization using a binary GA. The left graph shows the distribution of the population. The right graph shows the maximum and mean fitness of the population (the upper and lower plots respectively).

now used to determine the global maximum in the range with a population of 20, a mating proportion of 60% and a mutation rate of 5%. Each member of the population comprises 12 bits, and hence in the search range 0 to 1 the maximum accuracy is 0.0833. Figure 2.2 shows the progress of running the algorithm once. The left hand diagram of the figure shows the distribution of the population at every 10th generation. It can be seen that initially the population clusters around the local maximum value close to $x = 0.25$, but after approximately 50 generations the population clusters around the global maximum at $x = 0.08$. The maximum function and mean function values are shown in the right hand diagram of the figure. Initially, the maximum value of the function is about 0.95, but sometime after 50 generations changes to close to 1. Note that the mean value of the population fluctuates with the generation. After 20 runs, the maximum function value is 0.999988 at $x = 0.0796093$, an error in x of approximately 0.1%. The mean value of the 20 runs is 0.981143.

2.3 ALTERNATIVE METHODS OF CODING

In the previous section chromosomes were generated comprising a string of binary digits and these were interpreted as binary numbers and, when required, converted to decimal numbers. An alternative coding is to interpret the binary digits as a Gray code. Gray code is a binary number system where two successive numbers differ by only one binary digit or bit. The code was originally developed by Gray to make the operation of systems of mechanical switches more reliable.

A comparison of the three bit Gray code and the three bit binary code is shown below.

Decimal	0	1	2	3	4	5	6	7
Binary	000	001	010	011	100	101	110	111
Gray	000	001	011	010	110	111	101	100

Other forms of coding include octal coding, hexadecimal coding and decimal coding. Decimal coding is described in detail in Section 2.8.

2.4 ALTERNATIVE METHODS OF SELECTION FOR MATING

There are several alternatives to using a roulette wheel biased by fitness to select chromosomes for mating. Here we discuss three of them, although there are further variants within each of these methods.

Elitism: The biased random selection described in Section 2.2 does not guarantee that the fittest member or members of the population will be selected for mating. Thus the current best solution may be discarded! This will allow for more exploration, at the expense of convergence, which may lead to a better solution for a diverse function with many variables in a diverse region. At the expense of exploration and to speed up the algorithm, the fittest, or perhaps to the two fittest members of the population are automatically selected to proceed to the next generation without mating and without mutation. This is called elitism.

Roulette wheel selection by rank: We have described how a roulette wheel may be divided so that sectors are proportional to the fitness of the population members. This is shown in Figure 2.1 (a) and (b). If one or two members of the population are significantly fitter than the rest, they will have a very high probability of being chosen for mating to the detriment of diversity. It is always important to obtain a good balance between exploration and convergence. Dividing the roulette wheel in proportion to reverse rank overcomes the problem of a small number of very fit population members having a very high probability of being chosen. For example, a population of four members is ranked 1, 2, 3, 4, with proportions 0.60, 0.25, 0.10 and 0.05 on the roulette wheel. Clearly, one member of the population is very likely to be chosen. The reverse rank is 4, 3, 2, 1 and the roulette wheel is then divided in the proportions, 0.4, 0.3, 0.2 and 0.1, so the highest ranking member of the population is now less likely to be chosen (a 40% chance rather than a 60% chance). This is illustrated for the data of Table 2.1 and in Figure 2.1. In this particular example selection by rank is not significantly different from selection by fitness.

Tournament selection: In this procedure, a group of chromosomes is randomly selected. The fittest members of the group are then chosen for mating. This is deterministic

Table 2.4 Members of the population chosen for mating by tournament selection from randomly selected pairs

Ref	$g(\mathbf{x})$	Random selection	Random selection	Fittest of pair	$g(\mathbf{x})$
#0	0.1045	#0	#1	#1	0.1046
#1	0.1046	#9	#9	#9	0.5615
#2	0.3483	#5	#0	#5	0.4802
#3	0.2556	#2	#3	#2	0.3483
#4	0.7505	#8	#0	#8	0.9826
#5	0.4802	#0	#1	#1	0.1046
#6	0.9748	#6	#7	#6	0.9748
#7	0.3277	#6	#4	#6	0.9748
#8	0.9826	#5	#2	#5	0.4802
#9	0.5615	#7	#1	#7	0.3277
Total	4.8904				5.3393

tournament selection. Some researchers prefer to introduce a further random element into the tournament, so that the winner of each tournament is not necessarily the fitter, although the winner of the tournament is biased in favor of the fitter. Table 2.4 gives an example of a deterministic tournament selection. The initial population of 10 members and their individual fitnesses $g(\mathbf{x})$ are the same as that shown in Table 2.1. Consider line three of Table 2.4. This shows that population members #5 and #0 were randomly chosen, and because member #5 is fitter than member #0 (a fitness of 0.4802 compared with 0.1045), #5 is chosen to be part of the mating pool. Note that the total fitness of the mating pool, 5.3393, is higher that the total fitness of the original population, 4.8904. There are other variations of the tournament selection which we do not consider.

2.5 ALTERNATIVE FORMS OF MATING

In Section 2.2 chromosomes were mated by a randomly selected single point crossover. Shown below are two chromosomes before mating.

$$\text{Parent 1} \quad a_0 \quad a_1 \quad a_2 \quad a_3 \quad a_4 \quad a_5 \quad a_6 \quad a_7 \quad a_8 \quad a_9$$
$$\text{Parent 2} \quad b_0 \quad b_1 \quad b_2 \quad b_3 \quad b_4 \quad b_5 \quad b_6 \quad b_7 \quad b_8 \quad b_9$$

If the two chromosomes are mated by a randomly chosen crossover after the 4th gene then the result is as follows.

$$\text{Child 1} \quad a_0 \quad a_1 \quad a_2 \quad a_3 \quad b_4 \quad b_5 \quad b_6 \quad b_7 \quad b_8 \quad b_9$$
$$\text{Child 2} \quad b_0 \quad b_1 \quad b_2 \quad b_3 \quad a_4 \quad a_5 \quad a_6 \quad a_7 \quad a_8 \quad a_9$$

An alternative is to use a two point crossover. The result of a two point crossover after the 4th and 7th gene is shown below.

$$
\begin{array}{cccccccccc}
\text{Child 1} & a_0 & a_1 & a_2 & a_3 & b_4 & b_5 & b_6 & a_7 & a_8 & a_9 \\
\text{Child 2} & b_0 & b_1 & b_2 & b_3 & a_4 & a_5 & a_6 & b_7 & b_8 & b_9
\end{array}
$$

A further alternative is to use a randomly selected multi-crossover. In this procedure, each pair of corresponding genes are randomly chosen to either remain the same or interchange. In the example shown below 3rd, 7th, 8th and 9th gene pairs interchange.

$$
\begin{array}{cccccccccc}
\text{Child 1} & a_0 & a_1 & b_2 & a_3 & a_4 & a_5 & b_6 & b_7 & b_8 & a_9 \\
\text{Child 2} & b_0 & b_1 & a_2 & b_3 & b_4 & b_5 & a_6 & a_7 & a_8 & b_9
\end{array}
$$

There are alternative ways of mating that do not involve crossover. One such alternative is to use logical operators. Consider the nth binary digit in pair of parent chromosomes, $\mathbf{a}$ and $\mathbf{b}$ that are chosen to mate. The nth binary digit of the child is computed from the following equations, where r_n is a randomly generated binary digit, i.e. 0 or 1.

$$
c_n = (a_n \text{ and } b_n) \text{ or } (r_n \text{ and } (a_n \text{ xor } b_n)) \tag{2.1}
$$

Note that and, or and xor are logical bit-wise operations called and, or, and exclusive-or, that operate at the level of individual bits.

A second offspring can be created by simply using a different set of random binary digits. An example which implements the equation shown above is now given:

Parent 1	1	0	0	1	0	1	0	0	0	0	0	1	0	0
Parent 2	0	1	1	0	1	0	0	1	1	0	1	1	0	0
Random number	1	1	0	1	1	0	0	1	1	1	0	1	1	0
Child 1	1	1	0	1	1	0	0	1	1	0	0	1	0	0
Random number	0	0	0	1	0	0	1	0	0	0	0	0	1	1
Child 2	0	0	0	1	0	0	0	0	0	0	0	1	0	0

The decimal equivalent values of the two parents are $(0.3492, -2.9059)$ and $(-0.3492, -0.4588)$. Child1 and Child2 are $(1.4286, -0.6471)$ and $(-1.7460, -2.9059)$ respectively.

Whilst this method generates two children with an element of randomness, it is unclear how the desirable properties of the parent chromosomes are passed on to the their offspring.

2.6 ALTERNATIVE FORMS OF MUTATION

In Section 2.2 mutation is described as a process whereby a binary digit is flipped from 0 to 1 or vice versa. An alternative to this is to flip two digits. For example

| Original | 1 0 0 <u>1</u> 0 1 0 0 <u>0</u> 0 0 1 0 0 |
| Mutated | 1 0 0 <u>0</u> 0 1 0 0 <u>1</u> 0 0 1 0 0 |

A further alternative is to interchange the *position* of one or more digits. This is illustrated for a pair of digits below.

| Original | <u>1</u> <u>0</u> 0 1 0 1 0 0 0 0 <u>0</u> <u>1</u> 0 0 |
| Mutated | <u>0</u> <u>1</u> 0 0 0 1 0 0 1 0 <u>1</u> <u>0</u> 0 0 |

This shows that in contrast to flipping digits, the number of 1s and 0s in the mutated chromosome remain the same as the original one. This has advantages in certain classes of problems.

The mutation rate can remain constant from generation to generation, or it can be decreased with passing generations. This has the advantage that in early generations the mutation encourages new search areas, whereas in later generations focus remains on the emerging optimal solution.

2.7 THEORETICAL BACKGROUND TO GAS

Although GAs have been shown to work well in practice, attempts to establish an acceptable theoretical basis for their operation has proved difficult. A comprehensive theory of GAs should be able to explain why the GA is able to determine an optimal solution, even though only a very small percentage of the solution space is tested. It should also indicate which problems can be solved efficiently.

Schema theory (Holland, 1992) attempts to provide a theoretical basis for the GA. A schema (its plural is schemata) is important because it illustrates the degree to which two strings are similar and thus contain similar information. For example, consider the binary strings 101011 and 101101. Both are examples of the schema 101##1 where the symbol # indicates that at that position we can have either a 0 or 1. The order of the schema is the number of positions not defined by #. In this example the order of the schemata is 4.

The schema theorem suggests that schemata that are short and of low order are more likely to survive and have an average fitness greater than the average fitness of the whole population. In addition these schemata, called 'building blocks', are given an exponentially increasing number of trials as the process continues. Early workers believed that crossover was the source of efficiency in the GA, the effect of mutation was largely ignored. They describe the working of a GA as a process where by short, lower order, fit schema were selected, recombined with crossover to form strings of increasingly high fitness. This concept was called the building-block hypothesis. This was described by Goldberg (1989) as, given any low order short defining length schema partitions the GA is expected to converge to the schema in that partition with the highest average fitness.

Over the years schema theory has been criticized as being an inadequate theoretical basis for the GA (see Fogel and Ghozeil, 1997) and alternative proposals have been suggested. A full explanation of schema theory can be found in Coley (1999).

2.8 CONTINUOUS OR DECIMAL CODING

The Continuous or Decimal or Real-Coded Genetic Algorithm is similar in structure to the binary form of the Genetic Algorithm described in Section 2.2, in that an initial population in the region of interest is generated randomly, pairs are selected from the current population and are mated according to fitness, and mutation of chromosomes occurs with a specified probability. However, these steps have significant differences in their implementation in the continuous GA. Basing our description on an optimization problem, we randomly generate a set of chromosomes. This initial population is a set of randomly chosen real numbers rather than binary digits. The key feature here is the values can be any of the continuous set of values in the region of interest and not a discrete set of binary values that we have used in the binary form of the algorithm. Suppose we assume that the function to be optimized has four variables. Then initially each chromosome is a vector of four randomly generated decimal numbers, each lying in the search range for the variable. If we choose to have a population of 20 chromosomes, then each has its fitness calculated using a fitness function and a number of the fittest are chosen for the mating process. For example the 8 fittest chromosomes from a group of 20 chromosomes may be chosen to constitute the mating pool. From this group, random pairs are chosen for mating.

The mating process is again broadly similar to that of the binary form of the GA in that random pairs of chromosomes are selected for mating. However, for the continuous GA there are many procedures that may be used to create offspring, and combinations of procedures make the choice of mating methods almost endless! Here we introduce a selection of them.

Crossover. A point is randomly chosen in order that the parental chromosomes are intermixed about this point by simply interchanging the real variable values within the chromosomes. However, in contrast to the binary GA, where crossover generates new values, in the continuous GA crossover simply interchanges the original set of randomly generated real values without producing new values in the region. For example, in a problem with 5 variables consider two selected parents $\mathbf{p}_1$ and $\mathbf{p}_2$ shown below. Assume a two point crossover, where the randomly chosen crossover positions are after the second and fourth variable. Then the children vectors $\mathbf{c}_1$ and $\mathbf{c}_2$ are as follows.

Parents $\mathbf{p}_1 = [12.3, 7.1, 19.7, 4.3, 56.0]$ and $\mathbf{p}_2 = [15.8, 2.7, 11.4, 9.6, 23.7]$

Children $\mathbf{c}_1 = [12.3, 7.1, 11.4, 9.6, 56.0]$ and $\mathbf{c}_2 = [15.8, 2.7, 19.7, 4.3, 23.7]$

As can be seen the two children are different from their parents with different fitness values, but no new genetic data have been generated. This is in contrast with the situation that prevails after crossover in a binary GA.

Blending. To help explore the region we need to introduce new values and this can be achieved using various forms of blending. For example, suppose the function to be minimized is a function of n_{var} variables and the two chromosomes randomly chosen to be mated are $\mathbf{p}$ and $\mathbf{q}$. Then the children $\mathbf{c}$ and $\mathbf{d}$ are given by

$$c_k = r_u p_k + (1 - r_u)q_k, \quad d_k = (1 - r_u)p_k + r_u q_k, \quad k = 1, 2, ..., n_{var} \qquad (2.2)$$

where r_u is a random number taken from a uniform distribution in the range 0 to 1. Blending may be applied to every variable using the same value of r_u, i.e. $\mathbf{c} = r_u\mathbf{p} + (1 - r_u)\mathbf{q}$ and $\mathbf{d} = (1 - r_u)\mathbf{p} + r_u\mathbf{q}$. Alternatively, the blending may be applied to every variable using a *different* value of r_u, or it might be applied to only some of the variables. It can also be combined with crossover, where some variables are crossed over and some are blended. Note that a limitation of blending is that the children cannot be outside the hypercube defined by the two parents.

Extrapolation. The is a similar process to blending but it can generates new values of the variables outside of the hypercube defined by the two parents. Three children, $\mathbf{c}$, $\mathbf{d}$ and $\mathbf{e}$ are generated from two parents, $\mathbf{p}$ and $\mathbf{q}$ as follows:

$$c_k = 0.5p_k + 0.5q_k \qquad (2.3)$$
$$d_k = 1.5p_k - 0.5q_k \qquad (2.4)$$
$$e_k = -0.5p_k + 1.5q_k \qquad (2.5)$$

where $k = 1, 2 ..., v_{var}$. The best two children are chosen to proceed to the next generation. However, if one child is outside the bounds for which the variables are defined, it is discarded and the other two proceed to the next generation. Of course, a factor of 0.5 need not be chosen so that (2.3), (2.4), and (2.5) may be generalized by introducing a parameter λ in the range [0 1] where, in (2.3), (2.4), and (2.5), $\lambda = 0.5$. Thus,

$$c_k = \lambda p_k + (1 - \lambda)q_k \qquad (2.6)$$
$$d_k = (1 + \lambda)p_k - \lambda q_k \qquad (2.7)$$

Table 2.5 Children generated by extrapolation using (2.3), (2.4) and (2.5). Pairs of members of the population chosen randomly for mating. $\lambda = 0.5$

case	1	2		3	4		5	
p_k	7.2	7.2		5.8	1.2		8.4	
q_k	3.0	2.2		4.2	4.8		2.2	
c_k	5.1	4.7		5.0	3.0		5.3	
d_k	9.3	9.7		6.6	−0.6	u/s	11.5	u/s
e_k	0.9	−0.3	u/s	3.4	6.6		−0.9	u/s

$$e_k = -\lambda p_k + (1 + \lambda)q_k \tag{2.8}$$

Table 2.5 shows the result of computing the children c_k, d_k and e_k for particular values of a_k and b_k, using $\lambda = 0.5$. We assume that the variable the search is in the range 0 to 10. In cases 1 and 3, three valid children are generated, and the best two are chosen for the next generation. In cases 2 and 4, only two valid children are generated and are chosen for the next generation, in each case the third child is outside the search range. In case 5, only one of the children is valid, the other two are outside the search range, and so the children are discarded and the process repeated until at lease two valid children are generated.

The process of mutation is implemented in a very similar way to that of the binary GA. A mutation rate is chosen then the number of mutations can be calculated from the number of chromosomes and the number of components in the chromosome. Then positions are randomly selected in the chromosomes and these chromosome values are replaced by random values selected within the search region. This is another way of ensuring that the algorithm explores the whole region which increases the chance of finding the global minimum.

2.9 SELECTED NUMERICAL STUDIES USING THE CONTINUOUS GA

Test results given in this book are usually obtained by running the algorithm several times, in the examples here there are 20 runs of the continuous GA algorithm. For each run the best, that is the minimum function value of all values given by each member of the trial population is noted. Then, from the 20 runs, the best, the worst, the mean and standard deviations of the results are recorded and it is these results that are provided in the tables. Obviously, if the algorithm is only run once, the best, the worst and the mean would be identical and the standard deviation would be zero.

We now examine the effect of varying the mutation rate and population size when minimizing Rastrigin function with four variables, denoted by RAS4, using the continuous GA. Table 2.6 shows the influence of the mutation rate, r_m. In this example,

Table 2.6 Min of RAS4. $n_{gen} = 500$, 40 chromosomes, using 20 runs

mr	Mean	Best	Worst	St Dev
0.10	0.1998	4.3274×10^{-4}	2.2006	0.5453
0.05	0.0610	4.6546×10^{-6}	0.7717	0.1746
0.01	0.0174	5.5343×10^{-6}	0.1287	0.0372
0.001	0.0637	9.0810×10^{-5}	0.5168	0.1234

Table 2.7 Min of RAS4. $n_{gen} = 500$, mr = 0.01, 40 chromosomes, using 20 runs

n_{pop}	Mean	Best	Worst	St Dev
20	0.0468	1.9221×10^{-4}	0.2008	0.0633
40	0.0218	9.4457×10^{-5}	0.1800	0.0411
60	0.0241	3.8745×10^{-4}	0.1637	0.0437

Table 2.8 Minimization of RAS2, RAS4 and RAS6. 40 chromosomes, $r_m = 0.01$, $n_{gen} = 500$, using 20 runs

Function	Mean	Best	Worst	St Dev
RAS2	4.6070×10^{-4}	0	0.0036	9.2110×10^{-6}
RAS4	0.0559	8.9412×10^{-6}	0.4577	0.1335
RAS6	6.1165	1.6074	10.8329	2.5616

$r_m = 0.01$ appears to be best, but there is a random element in these results, so it is not certain that this will always be the case when optimizing this function. Of course, Rastrigin's function with more variables, or another problem altogether, may require a different best value of r_m.

Table 2.7 shows that the population size, n_{pop} is not significant when minimizing the four variable Rastrigin function with four variables but might be for a larger problem.

We now compare the performance of the continuous GA when minimizing Rastrigin's function with two, four and six variables, denoted by RAS2, RAS4 and RAS6 respectively. Table 2.8 shows that the performance of the continuous GA declines as the number of variables is increased.

Table 2.9 shows the effect of increasing the number of generations used. The function being optimized is the four variable Rastrigin function denoted RAS4. Clearly the more generations the better the result.

Figures 2.3 and 2.4 illustrate graphically the nature of convergence of the algorithm we consider the Styblinski and Tang and the Rastrigin functions in two variables. The graphs show the position of the points at 0, 10, 20 and 50 generations.

Table 2.9 Minimization of the function RAS4. Mutation rate set at 0.01 and the number of chromosomes at 40, using 20 runs

n_{gen}	Mean	Best	Worst	St Dev
500	0.0126	3.1587×10^{-4}	0.0970	0.0236
200	0.5050	0.0129	1.5831	0.5648
50	1.9880	0.38745	3.2532	0.9571

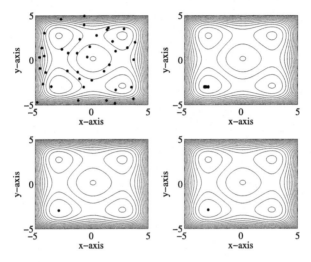

Figure 2.3 Contour plot of the Styblinski-Tang function showing the initial distribution of the population and the distribution after 10, 20 and 50 generations.

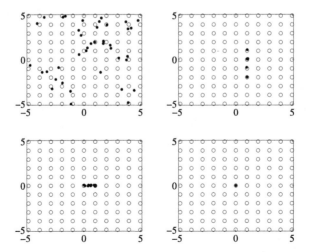

Figure 2.4 Plot showing the location of the minima of Rastrigin's function together with the initial distribution of the population and the distribution after 10, 20 and 50 generations.

2.10 SOME APPLICATIONS OF THE GENETIC ALGORITHM

Because the GA was one of the first metaheuristic optimization methods, it has been very widely used in many disciplines. For example, Rajpaul (2012) provides a review of some of the applications of GAs to astronomy and astrophysics. Among the applications of GAs he lists are: the task of determining the orbital parameters of interacting galaxies; the challenging task of finding the orbital parameters of planets orbiting a distant star based on radial velocity measurements; the reconstruction of the expansion history of the universe; performing fits to stellar spectra to allow many stellar properties to be inferred; and finally autonomous telescope scheduling to ensure the efficient achievement of many different scientific objectives.

Kiehbadroudinezhad et al. (2014) have used a GA to optimize the configuration of arrays of antenna used in radio astronomy. The observation of celestial objects necessitates a large array of antennae and optimizing this kind of array is very important in creating a high performance system.

A further example is the application of GAs to locate and possibly quantify damage in structures. When a structure is damaged its stiffness changes and this in turn changes the structure's vibration characteristics. Essentially damage can be located in a structure as follows. A mathematical model of the structure is created and used to predict the vibration characteristics of the structure for every possible damage scenario. The vibration characteristic of the damaged structure is measured and damage is located by minimizing the errors between the measured and predicted vibration characteristics. Friswell et al. (1998) used a binary GA to do this. In a more recent paper Meruane and Heylen (2011) use a real or continuous GA. In this later work, the problem had 43 dimensions.

We now describe in more detail the application of a genetic algorithm to the optimization of the design of a pin jointed steel frame structure with a minimum weight for a particular task, i.e. to support a particular distribution of loads. Engineers seek to minimize the weight of steel in order to minimize the cost of the steel, subject to the constraints that the stresses in the steel must not exceed the safe working stress and the deflections in the frame do not exceed a specified maximum.

A further complication is that the members of the framework must be constructed of standard steel sections, it would not be cost effective to have special sections rolled. Thus each member of the framework is chosen from a set of standard steel sections. Here we outline the approach of Rajeev and Krishnamoorthy (1992) to solve this problem using a GA. Many papers have been written on this topic, for example Camp et al. (1998).

Suppose the framework consists of n members, then the problem becomes

$$\text{minimize } f(\mathbf{x}) = \sum_{j=1}^{n} m_j = \sum_{j=1}^{n} \rho A_j L_j$$

where m_j is the mass of member j of the framework, A_j and L_j are the cross-sectional area and length of member j, respectively, and ρ is the density of steel. The density of steel and the length of the members is fixed, so only the cross sectional area of each member can be changed. The way in which the design variables $\mathbf{x}$ are related to the cross sectional areas of the members is described below.

The minimization problem is subject to the constraints

$$\sigma_j \leq \sigma_a, \ j = 1, 2, ..., n$$

where σ_j is the stress in member j, σ_a is the maximum allowable stress, and

$$u_k \leq u_a, \ k = 1, 2, ..., p$$

where u_k is a displacement in any of the directions x, y or z at a particular joint, u_a is the maximum allowable displacement in any direction, and p is the total number of displacements that we wish to limit to u_a. These constraints are satisfied if

$$g_j(\mathbf{x}) = \frac{\sigma_j}{\sigma_a} - 1 \leq 0, \ j = 1, 2, ...n$$

$$g_j(\mathbf{x}) = \frac{u_k}{u_a} - 1 \leq 0, \ j = n+1, n+2, ...n+p$$

To combine the effects of the constraints, a violation coefficient C is then defined as

$$C = \sum_{j=1}^{n+p} c_j$$

where

$$c_j = \begin{cases} g_j(\mathbf{x}), & \text{if } g_j(\mathbf{x}) > 0 \\ 0, & \text{otherwise} \end{cases} \tag{2.9}$$

Thus, the function to be minimized is

$$\phi(\mathbf{x}) = f(\mathbf{x})(1 + KC)$$

where the parameter K was chosen to be 20. Clearly, if one or more constraints are violated $\phi(\mathbf{x})$ will be large and we wish to minimize $\phi(\mathbf{x})$. However, the binary GA normally determines the maximum of a function. Since we wish to minimize $\phi(\mathbf{x})$ we maximize $F(\mathbf{x})$ where

$$F(\mathbf{x}) = [\phi(\mathbf{x})_{max} + \phi(\mathbf{x})_{min}] - \phi(\mathbf{x})$$

and $\phi(\mathbf{x})_{max}$ and $\phi(\mathbf{x})_{min}$ are the maximum and minimum values of $\phi(\mathbf{x})$ of the population respectively, determined at each generation. Thus the member of the population with the maximum value of $\phi(\mathbf{x})$ has the minimum value of $F(\mathbf{x})$ and vice versa. Since we can only choose from a set of beam sections, the design variables **x** are integer numbers and the binary GA lends itself to this type of problem. Suppose that it is decided to design from 16 possible standard sections. We can let the length of the chromosomes of the trial solutions be $4n$ where n is the number of members in the frame. Each member of the frame be represented by 4 binary digits, i.e. the integer numbers 0 to 15 in the trial solution. Thus each group of 4 binary digits a refer to one of the 16 possible sections. We can imagine the beam section properties in the form of a table. Each of the 16 rows of the table refer to a particular section and each column of the table list a section property: cross sectional area, position of the centroid, second moment of area, etc. In this particular problem we only require the cross sectional area. Thus a trial vector will generate the section row number and from it we can obtain the corresponding cross sectional area.

The complete process is summarized in the following steps:

Step 1: Randomly generate a population of trial chromosomes, each of $4n$ binary digits.

Step 2: For each member of the trial population we can determine the cross sectional area for each member of the frame and, knowing the lengths of the members and the density of steel, compute $f(\mathbf{x})$.

Step 3: Knowing the loads applied to the structure and the trial cross sectional areas of the members the equilibrium equations can be solved to determine the displacements of the joints and stresses in the members. Note that these equations must be solved for each trial solution.

Step 4: Knowing the displacements of the joints and stresses in the members we can determine which if any constraints are violated and from this determine C and hence $\phi(\mathbf{x})$. In turn we can obtain $F(\mathbf{x})$.

Step 5: Having the fitness for each member of the population, $F(\mathbf{x})$, the GA proceeds in the normal way, i.e., selection, mating and mutation to obtain a new and generally fitter trial population.

Step 6: Go to Step 2 and repeat the process until a satisfactory solution is obtained.

2.11 DIFFERENTIAL EVOLUTION

The Differential Evolution (DE) algorithm is a branch of evolutionary programming. It was developed by Storn and Price (1997) to solve difficult global optimization problems. The advantage of the DE algorithm is that it has a simple structure and is fast and robust. The DE algorithm uses mutation as the search mechanism and selection to direct the search towards a better solution. The principal difference between the DE algorithm and the GA is that the GA principally relies on crossover to obtain better solutions, whilst DE principally uses mutation as the primary search mechanism.

Consider a minimization problem in n_{var} dimensions. The search space is to be explored using n_{pop} trial solutions which are generated in the search space. The

DE algorithm requires column vectors of n_{var} real elements, denoted by $\mathbf{x}_i$, where $i = 1, 2, ..., n_{pop}$. In the terminology of DE the vectors $\mathbf{x}_i$ are called target vectors. This initial population is uniformly randomly distributed within the lower and upper bounds specified for each dimension. We now describe the major steps in this process.

Mutation: To evolve $\mathbf{x}_i^{(k+1)}$ for the $(k+1)$th generation we select from a uniform random distribution three target vectors from the current, kth, generation $\mathbf{x}_{r_1}^{(k)}$, $\mathbf{x}_{r_2}^{(k)}$ and $\mathbf{x}_{r_3}^{(k)}$ such that $r_1 \neq r_2 \neq r_3 \neq i$. The vector $\mathbf{v}_i^{(k+1)}$ is called the mutant vector and is defined as

$$\mathbf{v}_i^{(k+1)} = \mathbf{x}_{r_1}^{(k)} + F(\mathbf{x}_{r_2}^{(k)} - \mathbf{x}_{r_3}^{(k)}), \quad i = 1, 2, ... n_{pop} \tag{2.10}$$

where F is a user specified constant which controls the amplitude of the differential variation $(\mathbf{x}_{r_2}^{(k)} - \mathbf{x}_{r_3}^{(k)})$. The optimal range for this weighting or scaling factor F has been found to be in the range 0.5 to 1.0 although it can be in the range 0 to 2.

Crossover: The vector $\mathbf{u}_i^{(k+1)}$, called the candidate vector, is then obtained from

$$u_{ji}^{(k+1)} = \begin{cases} v_{ji}^{(k+1)} & \text{if } r_j \leq c_r \text{ or } j = s_i \\ x_{ji}^{(k)} & \text{if } r_j > c_r \text{ and } j \neq s_i \end{cases} \tag{2.11}$$

where $i = 1, 2, ... n_{pop}$, $j = 1, 2, ... n_{var}$, r_j is a random number chosen from a uniform distribution in the range 0 to 1, the crossover rate, c_r is a user supplied constant, also in the range 0 to 1, and s_i is a randomly chosen index in the range 1 to n_{var}. Thus (2.11) ensures that the candidate $\mathbf{u}_i^{(k+1)}$ will not be an exact copy of the target $\mathbf{x}_i^{(k)}$ and will get at least one parameter from the mutant vector $\mathbf{v}_i^{(k+1)}$.

Selection: The final step in deciding the values of $\mathbf{x}_i^{(k+1)}$ to become the new generation involves a selection process which for the minimization of the objective function $f(\mathbf{x})$ is

$$\mathbf{x}_i^{(k+1)} = \begin{cases} \mathbf{u}_i^{(k+1)} & \text{if } f(\mathbf{u}_i^{(k+1)}) \leq f(\mathbf{x}_i^{(k)}) \\ \mathbf{x}_i^{(k)} & \text{if } f(\mathbf{u}_i^{(k+1)}) > f(\mathbf{x}_i^{(k)}) \end{cases} \tag{2.12}$$

where $i = 1, 2, ... n_{pop}$. Using this new generation of trial solutions, the mutation, crossover and selection process is repeated until convergence is achieved. Note that the user is only required to specify three parameters, the number of trial solutions, n_{pop}, together with F and c_r.

It is possible that one or more of the new trial values might lie outside the search space in one or more dimensions. If that is the case, then one option is to simply take the trial value back to the boundary of the search space. Alternatively, we might move the trial value as far inside the boundary as it is currently outside the boundary. Thus, for the jth variable in the ith trial solution, x_{ji}, we apply the following adjustment.

$$\text{if } x_{ji} > x_{j(max)}, \text{ then } x_{ji} = x_{j(max)}$$

Table 2.10 Differential Evolution: randomly generated trial solutions

	$\mathbf{x}_1$	$\mathbf{x}_2$	$\mathbf{x}_3$	$\mathbf{x}_4$	$\mathbf{x}_5$	$\mathbf{x}_6$
	−1.7536	−2.5030	0.5523	2.4379	−1.7152	−1.3655
	2.1874	−2.4207	2.9430	2.6814	−1.7326	−0.2871
	2.5574	−0.5159	0.9284	1.5088	0.4470	0.5264
$f(\mathbf{x})$	14.4001	12.3910	9.8282	15.4097	6.1436	2.2241

Table 2.11 Showing $\mathbf{v}_1^{(1)} = \mathbf{x}_3^{(0)} + 0.8(\mathbf{x}_2^{(0)} - \mathbf{x}_4^{(0)})$, $\mathbf{v}_6^{(1)} = \mathbf{x}_4^{(0)} + 0.8(\mathbf{x}_2^{(0)} - \mathbf{x}_5^{(0)})$

$\mathbf{x}_3$	$\mathbf{x}_2$	$\mathbf{x}_4$	$\mathbf{v}_1$	$\mathbf{x}_4$	$\mathbf{x}_2$	$\mathbf{x}_5$	$\mathbf{v}_6$
0.5523	−2.5030	2.4379	−3.4005	2.4379	−2.5030	−1.7152	−1.3655
2.9430	−2.4207	2.6814	−1.1387	2.6814	−2.4207	−1.7326	2.1309
0.9284	−0.5159	1.5088	−0.6914	1.5088	−0.5159	0.4470	0.7384

Table 2.12 Crossover: generating $\mathbf{u}$ from $\mathbf{x}$ or $\mathbf{v}$

$\mathbf{x}_1$	$\mathbf{v}_1$	$r_1 > c_r$	$\mathbf{u}_1$	$\mathbf{x}_6$	$\mathbf{v}_6$	$r_6 > c_r$	$\mathbf{u}_6$
−3.4005	−1.7536	1	−1.7536	1.8076	−1.3655	1	−1.3655
−1.1387	2.1874	0	−1.1387	2.1309	−0.2871	0	2.1309
−0.6914	2.5574	0	−0.6914	0.7384	0.5264	0	0.7384
		$f(\mathbf{u}) \Rightarrow$	4.8498				6.9506

$$\text{if } x_{ji} > x_{j(min)}, \text{ then } x_{ji} = x_{j(min)}$$

or alternatively

$$\text{if } x_{ji} > x_{j(max)}, \text{ then } x_{ji} = 2x_{j(max)} - x_{ji}$$
$$\text{if } x_{ji} < x_{j(min)}, \text{ then } x_{ji} = 2x_{j(min)} - x_{ji}$$

The differential evolution procedure is now illustrated by a simple example. We seek to minimize the function $f(\mathbf{x}) = x_1^2 + x_2^2 + x_3^2$, limited in each variable to the range −3 to 3. We have chosen to make $c_r = 0.5$ and $F = 0.8$. We begin by randomly generating six trial solutions as shown in Table 2.10. The process for generating the next set of trial vectors (for $\mathbf{x}_1$ and $\mathbf{x}_6$ only) is now described.

Initially the parameters r_1, r_2 and r_3 are generated from a uniform random distribution of integers in the range 1 to n_{pop}. In this case when $i = 1$, $r_1 = 3$, $r_2 = 2$ and $r_3 = 4$ were randomly chosen; when $i = 6$, $r_1 = 4$, $r_2 = 2$ and $r_3 = 5$ were randomly chosen. Thus, substituting in (2.10) gives

$$\mathbf{v}_1^{(1)} = \mathbf{x}_3^{(0)} + 0.8(\mathbf{x}_2^{(0)} - \mathbf{x}_4^{(0)}), \quad \mathbf{v}_6^{(1)} = \mathbf{x}_4^{(0)} + 0.8(\mathbf{x}_2^{(0)} - \mathbf{x}_5^{(0)})$$

The calculations are shown in Table 2.11. We now compare $\mathbf{v}$ and $\mathbf{x}$, as shown in (2.11). In Table 2.12 a zero in the $r > c_r$ column indicates that the random number is

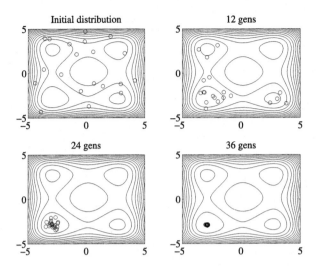

Figure 2.5 Contour plot of the Styblinski-Tang function showing the rapid convergence to the global minimum using Differential Evolution.

less than c_r, a one in the $r > c_r$ column indicates the random number was greater than c_r. It is seen that due to numbers chosen randomly, all elements of $\mathbf{u}_1$ are derived from $\mathbf{x}_1$. The first element of $\mathbf{u}_6$ is taken from $\mathbf{v}_6$, the others two elements are derived from $\mathbf{x}_6$. Having determined $\mathbf{u}_1$ and $\mathbf{u}_6$, we can compare $f(\mathbf{x}_1^{(0)})$ with $f(\mathbf{u}_1^{(0)})$, and $f(\mathbf{x}_6^{(0)})$ the $f(\mathbf{u}_6^{(0)})$. From Tables 2.10 and 2.12 we see that $f(\mathbf{x}_1^{(0)}) > f(\mathbf{u}_1^{(0)})$ and so from (2.12) we have $\mathbf{x}_1^{(1)} = \mathbf{u}_1^{(0)}$. Also from Tables 2.10 and 2.12 we see that $f(\mathbf{x}_6^{(0)}) < f(\mathbf{u}_6^{(0)})$ and hence from (2.12) we have $\mathbf{x}_6^{(1)} = \mathbf{x}_6^{(0)}$.

The ability of Differential Evolution to converge rapidly to a global minimum is illustrated in Figure 2.5. This shows the search for the global minimum of the Styblinski-Tang function. The randomly generated initial population of 20 members rapidly converges to the global minimum. After 36 generations it is clear that the population has migrated to the global minimum. More generations are required to obtain the value of the global minimum accurately.

2.12 OTHER VARIANTS OF DIFFERENTIAL EVOLUTION

The DE algorithm presented in the previous section is not the only form of differential evolution which has been shown to be of value. In order to classify the various forms of DE, Storn and Price (1997) introduced the general notation $DE/x/y/z$. Using this notation, the following variants of DE have been shown to be of value.

DE/rand/1/bin—the basic method described in the previous section.
DE/best/1/bin

DE/rand/1/exp
DE/best/1/exp
DE/rand/2/bin
DE/best/2/bin
DE/rand/2/exp
DE/best/2/exp
DE/rand-best/1/exp
DE/rand-to-best/1/bin

The interpretation of this notation is as follows. In the notation *DE/x/y/z*, *x* specifies the vector to be mutated, *y* is the number of vectors used and *z* specifies the form of crossover. Consider only the first part of *DE/x/y/z*, i.e. *DE/x/y/*. This part of the definition is described by the following equations:

DE/rand/1/ is described by (2.10)

DE/best/1/ is described by

$$\mathbf{v}_i^{(k+1)} = \mathbf{x}_{(best)}^{(k)} + F(\mathbf{x}_{r_1}^{(k)} - \mathbf{x}_{r_2}^{(k)}) \qquad (2.13)$$

DE/rand/2 is described by

$$\mathbf{v}_i^{(k+1)} = \mathbf{x}_{r1}^{(k)} + F_1(\mathbf{x}_{r_2}^{(k)} - \mathbf{x}_{r_3}^{(k)}) + F_2(\mathbf{x}_{r_4}^{(k)} - \mathbf{x}_{r_5}^{(k)}) \qquad (2.14)$$

Here we can choose to make $F_1 = F_2 = F$

DE/best/2/ is described by

$$\mathbf{v}_i^{(k+1)} = \mathbf{x}_{(best)}^{(k)} + F_1(\mathbf{x}_{r_1}^{(k)} - \mathbf{x}_{r_2}^{(k)}) + F_2(\mathbf{x}_{r_3}^{(k)} - \mathbf{x}_{r_4}^{(k)}) \qquad (2.15)$$

DE/rand-to-best/1/ is described by

$$\mathbf{v}_i^{(k+1)} = \mathbf{x}_{r_1}^{(k)} + F_1(\mathbf{x}_{(best)}^{(k)} - \mathbf{x}_{r_2}^{(k)}) + F_2(\mathbf{x}_{r_3}^{(k)} - \mathbf{x}_{r_4}^{(k)}) \qquad (2.16)$$

In (2.13) to (2.16), $i = 1, 2, \ldots n_{pop}$ and $r_1 \neq r_2 \neq r_3 \neq \ldots \neq i$.

Binomial crossover: Basically, a component of the offspring is taken with probability c_r from the mutant vector $\mathbf{u}_i^{(k+1)}$ and with probability $(1 - c_r)$ from the current element of the population $\mathbf{x}_i^{(k)}$, see (2.11).

Exponential crossover: Here the parameters of the trial vector are inherited from the corresponding mutant vector starting from a randomly chosen parameter index until the first time *rand* > c_r. The remaining parameters of the trial vector are copied from the target vector.

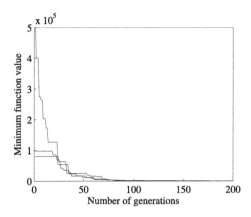

Figure 2.6 Minimization of Rosenbrock's function in 12 variables, with $F = 0.85$ and $c = 0.5$ and a population of 16. Three separate runs are shown.

Ali et al. (2011) have introduced the Centroid Differential Evolution (CDE), a variant of the basic DE. In the CDE method the equation

$$\mathbf{v}_i^{(k+1)} = (\mathbf{x}_{(best)}^{(k)} + \mathbf{x}_{r_1}^{(k)} + \mathbf{x}_{r_2}^{(k)})/3 + F(\mathbf{x}_{r_1}^{(k)} - \mathbf{x}_{r_2}^{(k)}), \quad i = 1, 2, \dots n_{pop} \tag{2.17}$$

$$\mathbf{v}_i^{(k+1)} = \begin{cases} (\mathbf{x}_{(best)}^{(k)} + \mathbf{x}_{r_1}^{(k)} + \mathbf{x}_{r_2}^{(k)})/3 + F(\mathbf{x}_{r_1}^{(k)} - \mathbf{x}_{r_2}^{(k)}) & \text{if } r_j \leq p_c \\ \mathbf{x}_{r_1}^{(k)} + F(\mathbf{x}_{r_2}^{(k)} - \mathbf{x}_{r_3}^{(k)}) & \text{if } r_j > p_c \end{cases} \tag{2.18}$$

These variations on the basic DE algorithm have been found to be of advantage when solving certain problems, but there is no universally accepted superior variant.

2.13 NUMERICAL STUDIES

Figures 2.6 and 2.7 show Differential Evolution being used to find the minimum of Rosenbrock's and Rastrigin's function respectively, each in 12 variables. Clearly Rosenbrock's function needs far fewer generations than Rastrigin's function to obtain a satisfactory result. We also see that each run provides different minimum values of the function after each generation, but ultimately they converge to essentially the same result.

Tables 2.13 and 2.14 illustrate the application of Differential Evolution in optimizing Rastrigin's function in 2, 4, 6, 8 10 and 12 variables. In Table 2.13 the number of generations employed is computed from $n_{gen} = 10 \times n_{var}^2$. This relationship is chosen to try to take account of the increasing difficulty in finding an optimal solution as the number of variables increases. However, the increase in difficulty is unlikely to be a simple square law. Table 2.13 shows that relatively good results are obtained for the minimization of Rastrigin's function for 4, 6, 8 and 10 variables but poor results for

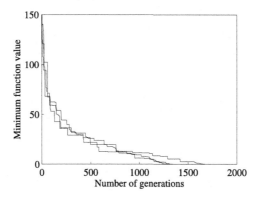

Figure 2.7 Minimization of Rastrigin's in 12 variables, with $F = 0.85$ and $c = 0.5$ and a population of 30. Three separate runs are shown.

Table 2.13 $F = 0.85, c = 0.50, n_{pop} = 30$, 20 runs

	n_{gen}	**Mean**	**Best**	**Worst**	**St Dev**
RAS02	40	1.5173×10^{-1}	3.4890×10^{-3}	9.4407×10^{-1}	2.2498×10^{-1}
RAS04	160	1.4744×10^{-4}	2.0169×10^{-6}	4.6987×10^{-4}	1.4970×10^{-4}
RAS06	360	6.2902×10^{-6}	9.0131×10^{-10}	9.7678×10^{-5}	2.1769×10^{-5}
RAS08	640	5.7811×10^{-6}	5.2664×10^{-9}	9.1767×10^{-5}	2.0284×10^{-5}
RAS10	1000	9.9544×10^{-3}	3.7396×10^{-9}	1.5397×10^{-1}	3.4237×10^{-2}
RAS12	1440	1.6771	1.4288×10^{-4}	7.4740	2.7125

Table 2.14 $F = 0.85, c = 0.50, n_{pop} = 30$, 20 runs

	n_{gen}	**Mean**	**Best**	**Worst**	**St Dev**
RAS02	60	2.8095×10^{-3}	5.2437×10^{-6}	2.5376×10^{-2}	5.6148×10^{-3}
RAS04	240	1.9202×10^{-9}	1.2951×10^{-11}	1.3504×10^{-8}	3.6286×10^{-9}
RAS06	540	6.0485×10^{-14}	0	5.4889×10^{-13}	1.2315×10^{-13}
RAS08	960	2.5757×10^{-15}	0	1.9540×10^{-14}	6.1663×10^{-15}
RAS10	1500	9.1882×10^{-13}	0	1.1969×10^{-11}	2.7549×10^{-12}
RAS12	2160	5.8069×10^{-9}	0	1.0952×10^{-7}	2.4434×10^{-8}

2 and 12 variables (confirming that the degree of difficulty of the problem does not follow a square law). In Table 2.14 the number of generations is doubled, compared with Table 2.13 and excellent results are obtained. In Table 2.15 the same problem is studied, but the number of generations is computed from $n_{gen} = 100n_{var}$. Good results are obtained for all except for the 12 variable case using 1200 generations. Doubling the number of generation gives much improved results. Table 2.16 shows the influence of the user chosen parameter F when optimizing Rastrigin's function with 6 variables. We see that for this problem $F = 0.6$ and $F = 0.85$ (the recommended value) give the best estimates of the minimum value of the function. Table 2.17 shows the influence of

Table 2.15 $F = 0.85, c = 0.50, n_{pop} = 30$, 20 runs

	n_{gen}	Mean	Best	Worst	St Dev
RAS02	200	0	0	0	0
RAS04	400	0	0	0	0
RAS06	600	0	0	0	0
RAS08	800	1.4446×10^{-10}	1.1013×10^{-13}	1.2251×10^{-9}	2.8438×10^{-10}
RAS10	1000	6.4591×10^{-4}	2.0715×10^{-7}	9.4032×10^{-3}	2.1313×10^{-3}
RAS12	1200	5.3691	1.5326×10^{-1}	9.4557	2.8945
RAS10	2000	0	0	0	0
RAS12	2400	5.5672×10^{-10}	0	1.1113×10^{-8}	2.4846×10^{-9}

Table 2.16 Optimization of RAS6. $c = 0.5, n_{pop} = 30, n_{gen} = 500$, 20 runs

F	Mean	Best	Worst	St Dev
0.30	7.1150×10^{-2}	0	9.9496×10^{-1}	2.3746×10^{-1}
0.60	2.3093×10^{-15}	0	4.6185×10^{-14}	1.0327×10^{-14}
0.85	3.6785×10^{-11}	2.3093×10^{-14}	4.5125×10^{-10}	1.1099×10^{-10}
1.20	6.3620×10^{-3}	1.8728×10^{-5}	1.1724×10^{-1}	2.6107×10^{-2}
1.50	3.6027	2.1660	6.0804	1.0378

Table 2.17 Optimization of RAS6. $F = 0.85, n_{pop} = 30, n_{gen} = 500$, 20 runs

c	Mean	Best	Worst	St Dev
0.50	4.5509×10^{-12}	1.7764×10^{-14}	3.2472×10^{-11}	8.9953×10^{-12}
0.60	1.6988×10^{-8}	4.5286×10^{-11}	1.4561×10^{-7}	3.5230×10^{-8}
0.70	8.0588×10^{-3}	9.8398×10^{-8}	1.4723×10^{-1}	3.2779×10^{-2}
0.80	1.5092	3.2936×10^{-3}	3.9651	1.3710
0.90	3.6641	3.2871×10^{-2}	8.1703	2.2872
1.00	10.4289	3.0632	30.1064	7.0487

Table 2.18 Optimization of RAS. $F = 0.85, c = 0.50, n_{pop} = 30, n_{var} = 6$, 20 runs. Algorithm 1 is $DE/rand/1/bin$ and Algorithm 2 is $DE/rand/2/bin$

fnc	Alg'm	n_{gen}	Mean	Best	Worst	St Dev
RAS6	2	500	6.74×10^{-5}	2.28×10^{-8}	6.25×10^{-4}	1.39×10^{-4}
RAS6	1	500	8.51×10^{-12}	0	1.13×10^{-10}	2.50×10^{-11}
RAS12	2	2000	4.644	1.952	7.097	1.493
RAS12	1	2000	4.98×10^{-2}	1.95×10^{-14}	9.95×10^{-1}	2.23×10^{-1}

varying the user chosen parameter c; the results confirm that for this problem, at least, the recommended value of c, i.e. $c = 0.5$ give best results.

Table 2.18 compares the performance of $DE/rand/1/bin$ with $DE/rand/2/bin$ in optimizing Rastrigin's function in 6 variables. For this particular problem $DE/rand/1/bin$ is more efficient.

Table 2.19 Optimization of ROS6. Pop = 30, 20 runs

n_{gen}	c	F	Mean	Best	Worst	St Dev
200	0.50	0.85	4.6441	1.9522	7.0974	1.4935
500	0.50	0.85	6.7429×10^{-5}	2.2833×10^{-8}	6.2530×10^{-4}	1.3861×10^{-4}
500	0.10	0.90	5.8859×10^{-6}	6.5558×10^{-8}	2.2955×10^{-5}	6.3031×10^{-6}

The final Table of this Chapter, Table 2.19 shows the optimization of Rosenbrock's function with 6 variables. The values of the parameters c and F are less critical than the number of generations used.

2.14 SOME APPLICATIONS OF DIFFERENTIAL EVOLUTION

Das and Suganthan (2011) present a detailed review of the basic concepts of DE and its major variants, its application to multi-objective, constrained, large scale, and uncertain optimization problems. They also review theoretical studies that have been conducted on the DE algorithm. Finally, they provide an overview of the significant engineering applications that have benefited from the powerful nature of DE.

Chauhan et al. (2009) have used differential evolution (DE) to train a wavelet neural network (WNN). They test the efficacy of the network on bankruptcy prediction data-sets for US and other national banks. Further, its efficacy is also tested on other benchmark data-sets such as the Wisconsin Breast Cancer Data. In the majority of these tests the DE trained networks outperform existing methods.

Differential evolution has been applied to the optimal design of shell-and-tube heat exchangers by Babua and Munawarb (2007). The main objective in any heat exchanger design is the estimation of the minimum heat transfer area required for a given heat duty, as it governs the overall cost of the heat exchanger. Many configurations are possible by varying the outer diameter, pitch, and length of the tubes, tube passes, baffle spacing, etc. Hence the design engineer needs an efficient strategy in searching for the global minimum. In this study DE was successfully applied to this optimal design problem and it was found that DE was significantly faster compared to GA and yielded the global optimum for a wide range of the key parameters.

Vasan and Simonovic (2010) describe the development and optimization of a computer model that involves the hydraulic simulation of water distribution networks. A model is formulated with the objective of minimizing cost and the DE method is applied to two benchmark water distribution system optimization problems—New York water supply system and Hanoi water distribution network. The study yielded promising results as compared with earlier studies in the literature. The results of the analysis demonstrate that DE can be considered as a potential alternative tool for economical and reliable water distribution.

An overview of the major applications areas of differential evolution is presented by Plagianakos et al. (2008). In particular they note the strengths of DE algorithms in tackling many difficult problems from diverse scientific areas, including single and multi-objective function optimization, neural network training, clustering, and DNA micro-array (or biochip) classification. To improve the speed and performance of the algorithm they employ distributed computing architectures and demonstrate how parallel, multi-population DE architectures can be utilized in single and multi-objective optimization. Using data mining they present a methodology that allows the simultaneous discovery of multiple local and global minimizers of an objective function.

A widely used model in the field of hysteretic or memory-dependent vibrations is that of Bouc and Wen. Different parameter values extend its use to various areas of mechanical vibrations. Kyprianou et al. (2001) present some results using DE in Bouc–Wen model identification, using both simulated data and experimental data obtained from a nuclear power plant.

Parameter estimation for chaotic systems is an important issue in nonlinear science and has attracted increasing interests from various research fields, which could be essentially formulated as a multidimensional optimization problem. Peng et al. (2009) use DE to estimate parameters of the Lorenz chaotic systems.

2.15 SUMMARY

In this Chapter we have introduced the binary and the decimal, real valued or continuous Genetic Algorithm (GA). We have studied the performance of the continuous GA on a number of test problems. Generally these methods are slow and require many generations to obtain a satisfactory solution. We have also introduced the Differential Evolution (DE) algorithm and discussed its variations and tested it on a number of functions. This algorithm is simpler and faster than the genetic algorithm. We have given graphical illustrations where appropriate.

2.16 PROBLEMS

2.1 For $x = 10 : 1 : 20$, convert x to binary code. Then convert this binary code to Gray code.

2.2 For $f(x) = 3x^2 - 32x + 100$, find the values of $f(x)$ for $x = 0 : 1 : 7$, and arrange the binary values ascending values of $f(x)$. Choose two pairs and enter each pair into a tournament selection process to obtain new offspring. Calculate the new objective function.

2.3 The function $f(x, y) = (x^3 + 3)y \sin(5y) + 700$ is to be minimized using the real or continuous GA. The search range is $x = 0$ to $x = 7.5$ and $y = 0$ to $y = 5$. The initial trial population is

$$
\begin{array}{ccccccc}
x & 1.1 & 2.1 & 3.4 & 4.9 & 0.7 & 1.9 \\
y & 4.8 & 2.2 & 0.7 & 1.4 & 3.7 & 2.1
\end{array}
$$

Choose the two fittest member of the population and mate them by each of the following methods:
(1) crossover,
(2) using (2.2) with $r = 0.3$,
(3) using (2.3), (2.4) and (2.5).
In each case determine the new function values. Are the new members of the population fitter than their parents? Remember in case (3) one child must be rejected according to the guidelines given in the text.

2.4 We wish to find the minimum of the function $f(x) = \sqrt{x} \sin(x)$ in the search range 0 to 10 using differential evolution. The initial (randomly) chosen positions of the six trial solutions are 8.1, 1.3, 6.3, 2.8, 9.6, and 1.6. Assume $F = 1$ and $c_r = 0.5$. Compute one generation of the DE algorithm. If any trial solution falls outside the search range, then place it inside the search range so that it is as far inside the boundary as it was outside. For example if $x = -2$ it becomes $x = 2$ and if $x = 13$ then it becomes $x = 7$.
The integer random numbers generated to choose values of r_1, r_2 and r_3 (see (2.10)) are as follows.

i	r_1	r_2	r_3
1	5	2	6
2	6	4	3
3	1	6	5
4	6	5	1
5	4	1	3
6	1	5	4

The six random numbers required to evaluate (2.11) for x_1, ..., x_6 are as follows: 0.79, 0.95, 0.65, 0.04, 0.35 and 0.93.

CHAPTER 3

Particle Swarm Optimization Algorithms

3.1 ORIGINS OF PARTICLE SWARM OPTIMIZATION

This section provides a general outline of how the concepts of particle swarming can be translated into practical algorithms. This type of algorithm arose from the work of Kennedy and Eberhart (1995) which was related to the flocking of birds and the behavior of fish schools. The word boids is frequently used to represent flocking creatures and arises from a shortening of the term birdoid objects. We will briefly discuss the nature of flocking algorithms which simulate flocking behavior. Using these concepts we describe the development of the particle swarm optimization (PSO) method.

The key features of flocking are that each boid has a specific location and velocity and pursues the following aims:

Alignment. A boid will attempt to align itself with other boids by matching their velocity.
Cohesion. The individual boid will tend to steer towards the center of mass of the flock.
Separation. Clearly each boid attempts to avoid bumping into to its near neighbors by keeping a reasonable distance apart, sufficient for collisions to be avoided while not losing sight of the flock.

To implement these concepts as an algorithm we generate, for each boid, an initial set of locations and velocities randomly then the continuous updating of the values of velocity and location is achieved using relatively simple formula that reflects the properties of alignment, cohesion and separation. The result is the simulation of boid flocking over a defined area which enables the flock to explore the region they occupy in an efficient and cohesive way. A real life example of this area of research is work done in Oxford University on the flight of ibis flocks which fly in a Vee formation. Each member of the flock takes it in turn to lead. The ibis work in pairs to achieve an efficient, perhaps optimum, distribution of effort during the flight which leads to a Vee alignment as the optimal choice.

It was noted by Kennedy and Eberhart (1995) that these concepts could be applied to non-linear optimization problems and in particular to the class of very difficult problems where multiple optima are present and where it is important to obtain the global optimum for the problem: clearly the roaming nature of the boid flocking process will

Introduction to Nature-Inspired Optimization
DOI: 10.1016/B978-0-12-803636-5.00003-7
49

help in the exploration phase of the optimization process which aims to avoid being trapped at local optimum.

The basis of the method of applying these concepts to non-linear optimization problems is achieved by considering a range of candidate solutions for the optima which are the locations of boids within the flock or as a general concept the location of particles in a swarm. These candidate solutions are updated and move around the search region randomly, thus exploring the region for improved solutions which provide values for the optimum of the objective or fitness function. An important feature of these algorithms is that not only may the local optima be located but the global optima can be obtained. Indeed this is usually the main aim, unlike gradient methods where only local optima may be found. In the PSO algorithm, the updating formulae used reflect the flocking behavior. These formulae take account not only of the behavior of individual particles in relation to their near neighbors, but also the global behavior of the group particles as a whole.

The key features which control the efficiency of the algorithm are how thoroughly the region is explored and how the accuracy of solutions is improved and the balance between the two. This process is sometimes called exploration and exploitation.

3.2 THE PSO ALGORITHM

The key properties of the particles of the PSO algorithm are position and velocity. The position of a population of particles is randomly chosen within the boundaries of the search region and the value of the objective function calculated for each particle position. The initial values of the velocities are taken as zero or selected randomly within specific limits. Thus if v_i are the velocities of the i particles then a possible initial setting is:

$$\mathbf{v}_i^{(0)} = 0 \text{ for } i = 1, 2, ..., n_{pop} \tag{3.1}$$

where n_{pop} is the size of the population.

There are many alternatives for assigning the initial velocities. Appropriate formula is then used to update the velocity of each particle. Using these values of velocity, the positions of the particles are updated. Then it is determined if improvements to the objective function have been achieved. If improvements are obtained the values of the positions of the particles are accepted as the new positions. In addition the global best position for all particles of the swarm is updated. This process is repeated until a preset convergence criterion is met. Then the process is stopped and the solutions provided. We now give the specific formula for updating the values of velocity and position.

The velocity of each particle is calculated from the following equation at time t; in practice the t values are equivalent to the iteration or generation number of the process, starting with $t = 0$:

$$\mathbf{v}_i^{(t+1)} = \mathbf{v}_i^{(t)} + a\mathbf{r}_1 \circ [\mathbf{x}_{lopt}^{(t)} - \mathbf{x}_i^{(t)}] + b\mathbf{r}_2 \circ [\mathbf{x}_{gopt}^{(t)} - \mathbf{x}_i^{(t)}] \qquad (3.2)$$

where the values a and b are in general positive constants set by the user, usually at the start of the process. These values must be carefully selected since they have a significant influence on the success of the process and the nature of convergence of the method giving proper emphasis to the local and global search aspects of the algorithm. The $\mathbf{x}_{lopt}^{(t)}$ vector is the best position vector for the particles calculated using the objective function $f(\mathbf{x}_i)$ in the local region. The $\mathbf{x}_{gopt}^{(t)}$ vector is the current global best position vector. This vector is continuously updated at each iteration and will provide the final optimum location vector. The $\mathbf{v}_i^{(t)}$ and $\mathbf{x}_i^{(t)}$ are the current values of the velocity and position vectors. The values $\mathbf{r}_1$, $\mathbf{r}_2$ are selected from the uniform random distribution vector $\mathbf{r}_u$ in the range 0 to 1 and they are reselected at each step of the algorithm. Note that $\circ$ in (3.2) denotes element by element multiplication, or the Hadamard or Schur multiplication. For example if there $\mathbf{a}$ and $\mathbf{b}$ are vectors of two elements then

$$\mathbf{a} \circ \mathbf{b} = \begin{bmatrix} a_1 \\ a_2 \end{bmatrix} \circ \begin{bmatrix} b_1 \\ b_2 \end{bmatrix} = \begin{bmatrix} a_1 b_1 \\ a_2 b_2 \end{bmatrix}$$

It is important to emphasize that the last two terms of the velocity adjustment equation, (3.2), reflect the two competing features of the algorithm. Firstly to ensure sufficient exploration of the local region to find an acceptably accurate value of the local minimum. Secondly to ensure that the whole region is explored so that a global minimum can be found and consequently avoid getting stuck at a local minimum. Randomness plays a key role in this process. Thus the choice of a and b in (3.2) is crucial in ensuring the compatibility of these two aims and their selection is not a simple matter.

Once the velocity values are updated the new position values can be calculated from:

$$\mathbf{x}_i^{(t+1)} = \mathbf{x}_i^{(t)} + \mathbf{v}_i^{(t+1)} \qquad (3.3)$$

This equation drives the process on from point to point, and the new point can then be tested for improvement. Large values of velocity mean the region is explored rapidly but perhaps missing key optima. If changes to $\mathbf{x}_i$ are too small the progress of the algorithm may be very slow. Careful choice of parameters can ensure balanced progress. We now describe the basic steps that are used in the PSO method.

1. Initialize values for velocity variables. Define the constants a and b. Generate randomly an initial population of $\mathbf{x}_i$ values confined to the region of interest. Thus

$$\mathbf{x}_i = \mathbf{x}_{lo} + \mathbf{r}_u \circ (\mathbf{x}_{hi} - \mathbf{x}_{lo})$$

ensures $\mathbf{x}_i$ is in range $\mathbf{x}_{lo}$ to $\mathbf{x}_{hi}$. For the $\mathbf{x}_i$ values, obtain the corresponding objective function values and select the best value of the objective function.

2. Calculate $\mathbf{x}_{lopt}^{(t)}$ and $\mathbf{x}_{gopt}^{(t)}$. Then update the velocity values for each particle using (3.2)
3. Update location values for each particle using (3.3).
4. Calculate the new objective function values and calculate the best values for individual particles and the global best value.
5. Repeat the steps from step 2 until convergence has been achieved.

This completes the algorithm description.

Here we provide MATLAB like pseudo-code for showing the details of the implementation of the calculation of the values $\mathbf{x}_{lopt}^{(t)}$ and $\mathbf{x}_{gopt}^{(t)}$. This assumes an objective function to be minimized has been defined as $f(\mathbf{x})$. For an each value of i.

Algorithm 1 Current Local Best Values.

---
if $f(x(i,:)) < f(xlopt(i,:))$ **then**
  $xlopt(i,:) = x(i,:)$
end if
if $f(x(i,:)) < f(xgopt(i,:))$ **then**
  $xgopt(:) = x(i,:)$
end if

This algorithm is called the global best particle swarm algorithm because it uses in the update equations the current best overall optima of all the particles i.e. $\mathbf{x}_{gopt}^{(t)}$. A particle swarm algorithm very similar to this one has been developed in parallel with this called the local best particle swarm algorithm.

The basic algorithm we have described has been tested extensively by many researchers on a range of test problems and on many industrial applications and found to be generally successful. However some researchers noted certain shortcomings in the behavior of the algorithm and consequently many suggestions have been made for its improvement. Some of these are described in the next section.

3.3 DEVELOPMENTS OF THE PSO ALGORITHM

Clearly the algorithm has a relatively simple structure but its efficiency depends crucially on the choice of parameters a and b. These values determine the balance between global and local exploration of the region. There clearly is the possibility of an over rapid exploration of the region or too slow an exploration of the region. A very useful and comprehensive review of modifications to the basic PSO algorithm is provided by Engelbrecht (2005). The velocity values drive the rate of exploration and one early change to the structure of the algorithm was to modify the velocity factor by introducing an inertial weight to allow the user to modify the rate and nature of exploration of the search region. We designate w as the inertial weight.

Then we modify the velocity updating formula as follows:

$$\mathbf{v}_i^{(t+1)} = w\mathbf{v}_i^{(t)} + a\mathbf{r}_1 \circ [\mathbf{x}_{lopt}^{(t)} - \mathbf{x}_i^{(t)}] + b\mathbf{r}_2 \circ [\mathbf{x}_{gopt}^{(t)} - \mathbf{x}_i^{(t)}] \tag{3.4}$$

With this relatively simple modification we can adjust the value of w until we are satisfied with the convergence of the algorithm. Larger values of w will promote global exploration of the given region since this will produce a larger increase in the velocity value, whereas smaller values of w, where $0 < w < 1$, will promote the effective and accurate evaluation of the local region. For a detailed explanation of this approach see Chatterje and Siarry (2006). Since this presents us with the problem of selecting the parameter w, a further suggestion for improving the PSO algorithm is that the inertial weight is changed as the algorithm proceeds. Thus we need only initially provide a broad general range of values for the w parameter. The first method we discuss (Eberhart and Shi, 2000) provides the means of gradually adjusting w so the effect of the first term decreases as the optimization proceeds. Thus at iteration t we use the formula:

$$w^{(t)} = (w^{(0)} - w^{(t_{max})})\frac{(t_{max} - t)}{t_{max}} + w^{(t_{max})} \tag{3.5}$$

which provides a linear decreasing value for w. The value of t_{max} is the preset maximum number of iterations and $w^{(0)}$ and $w^{(t_{max})}$ are the initial and final values of w. For example, $w^{(0)} = 0.9$, $w^{(t_{max})} = 0.1$.

The second method modifies the value of w based on a measure of the relative improvement in the function values as approximations to the optimum value are calculated, see Eberhart and Shi (2000). A relative improvement factor $K^{(t)}$ is calculated at each iteration t where:

$$K_i^{(t)} = \frac{f(\mathbf{x}_{gopt}^{(t)}) - f(\mathbf{x}_i^{(t)})}{f(\mathbf{x}_{gopt}^{(t)}) + f(\mathbf{x}_i^{(t)})} \tag{3.6}$$

Then the weights are determined from:

$$w_i^{(t+1)} = w^{(0)} + (w^{(t_{max})} - w^{(0)})\frac{e^{K_i^{(t)}} - 1}{e^{K_i^{(t)}} + 1} \tag{3.7}$$

We note that as the difference between the current function value and the global optimum value becomes smaller on convergence, thus the value of K_i tends to zero hence the exponential value of K_i tends to one. Consequently the second term of (3.7) tends to zero and the weight values change less and less.

Another issue that arises and is related to values of the velocity becoming too high and the particles rapidly diverging. To avoid this the idea of velocity clamping was introduced. One way is to use a constriction coefficient (C) which was introduced as a multiplier of the right hand side of the velocity equation, Clerc and Kennedy (2002). This is calculated from

$$C = \frac{2k}{|2 - \phi - \sqrt{\phi(\phi - 4)}|} \tag{3.8}$$

where $\phi = ar_1 + br_2$. Consequently the velocity equation is modified to include this factor as follows

$$\mathbf{v}_i^{(t+1)} = C\left(\mathbf{v}_i^{(t)} + a\mathbf{r}_1 \circ [\mathbf{x}_{lopt}^{(t)} - \mathbf{x}_i^{(t)}] + b\mathbf{r}_2 \circ [\mathbf{x}_{gopt}^{(t)} - \mathbf{x}_i^{(t)}]\right) \tag{3.9}$$

Using (3.9) which depends on ϕ, if $\phi >= 4$ and $k \in [0, 1]$ then Clerc and Kennedy (2002) show the swarm is guaranteed to converge. The value k controls the nature of the exploration process. If k is small then fast local convergence is achieved if k is large then the region is more thoroughly explored.

Another way to restrict the rate of global exploration of the region by the particles is to introduce a limit on the velocity directly. This is fairly easy to implement and is often used. We consider the velocity components for each particle and impose the following simple restrictions:

$$v_{ij} = \begin{cases} v_{ij} & \text{if } v_{ij} < v_{max,j} \\ v_{max,j} & \text{otherwise} \end{cases} \tag{3.10}$$

The initial values of the $v_{max,j}$ are set arbitrarily by the user, or set to the difference between the maximum and minimum values of the positions of the particle in the jth dimension. Similarly a slow exploration of the region can be avoided by imposing a minimum value of the velocity which it should not fall below. For a detailed explanation of this procedure see Shahzad et al. (2009) and also Ghalia (2008). Clearly the choice of $\mathbf{v}_{max}$ is important and various suggestions have been proposed for its value which may be problem dependent.

The danger of this procedure is that all values may be assigned the same maximum value or the same minimum value and consequently get stuck at these points. Modifications have been introduced to this clamping method that allow the maximum value of the velocity to be changed as the algorithm proceeds. A relatively simple approach is to adjust the clamping velocity according to the following equation:

$$\mathbf{v}_{max}^{(t+1)} = \left(1 - \left(\frac{t}{t_{max}}\right)^p\right)\mathbf{v}_{max}^{(t)} \tag{3.11}$$

Depending on the value of p this will reduce the value of $\mathbf{v}_{max}^{(t)}$ at different rates so taking $p = 1$ provides a linear rate of reduction and as the algorithm proceeds t tends to t_{max} and

$\mathbf{v}_{max}$ tends to zero. The argument being that as we approach the optimum only small changes in velocity are required. Larger values of p will provide more gradual changes in v_{max}. This approach was suggested by Fan (2002). An alternative approach reduces v_{max} according to its current performance in obtaining reductions in the optimum value. Thus if there is no reduction in the current best value after a number of iterations then the value of $\mathbf{v}_{max}$ is changed. This was suggested by Schutte and Groenwold (2003). Additional control parameters have been proposed to avoid premature convergence, see Van den Berg and Engelbrecht (2002).

A further suggested improvement on the basic PSO method is to consider the problem of the selection of good values for the quantities a and b. This can be done on an experimental basis by checking the effect on convergence of different combinations of values of a and b, but this is time consuming. Alternatively these coefficients may be adjusted from iteration to iteration. This method was suggested by Ratnaweera et al. (2004). They studied the use of the formulae:

$$a = (a_{min} - a_{max})\frac{t}{t_{max}} + a_{max} \tag{3.12}$$

$$b = (b_{min} - b_{max})\frac{t}{t_{max}} + b_{max} \tag{3.13}$$

Here b_{max} and b_{min} are values selected to provide a bound on the values that b can take. Similarly a_{max} and a_{min} are values selected to give an appropriate range of values for a. Clearly as the process continues, t, the current iteration number, will approach t_{max}. This is defined as the maximum number of iterations allowed for the execution of the algorithm in a specific case and thus b will tend to b_{min} and similarly a will tend to a_{min} as the iterations proceed.

One major problem that occurs with the PSO method is that of premature convergence in this case a solution is reached but is not the true global optimum. To deal with this important problem a major alteration to the original PSO algorithm has been suggested, called the guaranteed convergence PSO or GCPSO.

The guaranteed convergence PSO (GCPSO) method requires updating formulae for both the position and velocity. However, the updating formulae have major differences when compared with the original PSO updating formulae and for a detailed description of the method see Van den Berg (2006). The key feature of this modification is that adjustments are made which by random variation provoke further search in the region of an optimal location to avoid premature convergence. It should be noted that this modification of the algorithm may not lead to faster convergence but may improve the consistency of the performance of the algorithm over the whole range of problems.

3.4 SELECTED NUMERICAL STUDIES USING PSO

The range of modifications of the basic PSO method we have described clearly need to be examined to see how effective they are, indeed some are alternative ways of doing the same thing.

We now provide some practical insight into the nature of the PSO by using it to solve some standard non-linear optimization test problems. It should be emphasized that these studies are intended to be illustrative rather than exhaustive. Studies are now undertaken on three test problems illustrating specific features of optimization. The first problem, Rosenbrock's function which has only one local minimum, the second problem due to Styblinski and Tang has a small number of local optima and a specific global minimum the third, Rastrigin's function has many local minima and consequently presents a harder test for finding the global minima. None of these problems demand excessive computer time for low dimensional problems and consequently may be easily replicated by the reader.

Rosenbrock's function is a classic test problem for gradient optimization methods and because the single optimum is located in a long shallow curved valley, it presents significant difficulties and takes many iterations to locate the optimum. It is used in these tests to show that the methods can also solve difficult single optima problems. The minimum value is $f(x) = 0$ and is obtained when $x_i = 1$ for $i = 1, 2$. The problem is usually defined for test purposes in the range $[-5, 5]$ for each variable, although the range $[-5, 10]$ is sometimes used.

The Styblinski-Tang function for two variables is not a demanding problem but has various local minima so it is a useful test to see that the global minimum is found at the point $x_1 = -2.903534$ and $x_2 = -2.903534$. With global minimum at -78.3323. (See Figure 3.6.) The problem is usually tested in the range of values for the independent variables x_1 and x_2 in the range $[-5, 5]$. This problem can easily extended to many dimensions.

The final problem we can consider in this section is Rastrigin's function which is a more demanding problem since it has many closely packed local minima of similar value and a significant danger of an optimization method becoming trapped in a particular local minimum rather than finding the global minimum. This has a global minimum at $\mathbf{x} = [0, 0]^\top$ with $f(x) = 0$. The problem is usually tested in the range $[-5, 5]$ for the independent variables.

We now use a modified version of the PSO algorithm and test it on the functions we have described in various ways. Figure 3.1 shows the convergence for the much more difficult problem of minimizing the Rastrigin function in 6 dimensions or variables. It shows a gradual reduction in the function value over the 2000 iterations and appears to indicate periods where little change occurs in the current best value of the objective function as the area is searched for improvements.

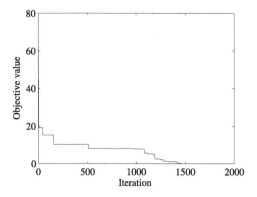

Figure 3.1 Minimizing Rastrigin's function with 6 variables, showing progress of convergence.

Table 3.1   Effect of number of iterations on the estimate of the minimum value of the function. $a = 2.05$ and $b = 2.05$

Iterations	ROS2	S-T2	RAS2	RAS4
200	3.3629×10^{-4}	-78.3323	1.1267×10^{-9}	2.2052
400	4.4627×10^{-7}	-78.3323	0	0.020526

Table 3.2   Effect of swarm size on the estimate of the minimum value of the function after 200 iterations. $a = 2.05$ and $b = 2.05$

Swarm Size	ROS2	S-T2	RAS2	RAS4
20	3.3629×10^{-4}	-78.3323	1.1267×10^{-9}	2.2052
10	6.3392×10^{-4}	-78.3323	5.7565×10^{-7}	2.2799

The most basic features that effect the performance of the algorithm are the number of iterations and the size of the swarm. In Table 3.1 we give some results for the test problems that show the difference between using 200 iterations and 400 iterations. In Tables 3.1 and 3.2, ROS2 indicates Rosenbrock's function with two variables, S-T2 indicates the Styblinski–Tang function in two variables and RAS2 and RAS4 indicates Rastrigin's function in two and four variables, respectively.

Recalling that the solution of the Rosenbrock and the two and four variable Rastrigin functions are zero and the solution of the Styblinski and Tang is -78.3323, Table 3.1 shows small or no improvements in the minimum values except for the last problem which is a more demanding one with four variables and shows a very large improvement with 400 iterations.

Similarly the effect of swarm size can be studied. Various researchers have recommended different swarm sizes, usually 10, 20 or 30 but sometimes more, depending on the problem. Swarm size is very important parameter since too few members of the swarm give poor results rapidly but too many members generally give more accu-

Table 3.3 Minimization of the function RAS4. 800 iterations

Swarm Size	Mean	Best	Worse	St Dev
10	0.5972	0	1.9899	0.7500
20	0.3447	0	5.8967	1.3255
30	1.0658×10^{-15}	0	2.1361×10^{-14}	4.7665×10^{-13}

Table 3.4 Minimization of the function RAS4. 2000 iterations

Swarm Size	Mean	Best	Worst	St Dev
10	0.0497	0	0.9950	0.2225
20	2.5743×10^{-12}	0	5.1486×10^{-11}	1.1513×10^{-11}
30	0	0	0	0

Table 3.5 Minimization of the function RAS6. 2000 iterations

Swarm Size	Mean	Best	Worst	St Dev
10	1.3318	0	37479	1.0550
20	0.5472	0	1.9899	0.7553
30	0.1996	0	0.9950	0.4080

rate results more slowly. Indeed a very large swarm size amounts to little more than enumeration of the possible solutions and a complete loss of efficiency.

Table 3.2 sets the number of iterations at 200 and illustrates the effect of different swarm sizes. The table shows only slightly less accurate results for the smaller swarm size for those problems which are not demanding.

A more thorough test is to perform many runs of the algorithm and take the mean of the results, thus smoothing out some of the effects of randomness. The results shown in Table 3.3 and 3.4 for the minimization of the RAS4 test function, using 20 runs. Two tests are performed, one with 800 iterations and one with 2000 iterations. Since the minimum value of this function is zero, it is clear that the larger number of iterations improves the accuracy of the solution and the larger the swarm size the better the optimum obtained for this particular function. To provide further information about the performance of the method we have included the best and worst results, together with the mean and the standard deviation of these results.

As a further illustration we minimize the Rastrigin function but with six variables (RAS6), a much harder problem. In this set of runs we have used 2000 iterations and swarm sizes 10, 20 and 30. Taking 20 runs gives the results shown in Table 3.5. Contrast these results with those of Table 3.4 which illustrates how the algorithm performs on the same function but with only four variables rather than six. The results shown in Table 3.4 are clearly much better and illustrate how increasing the number of variables in

Table 3.6 Effect of parameters *a* and *b* on optimization of RAS4. 800 iterations

Set	Mean	Best	Worst	St Dev
Set 1	0.0565	0	0.9950	0.2230
Set 2	0.2985	0	0.9950	0.4678

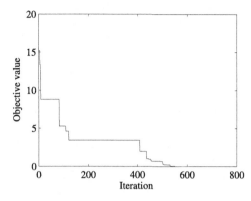

Figure 3.2 Changes in the objective function value for parameter set 1.

a problem disproportionately increases the problems difficulty. This is frequently called the "curse of dimensionality".

We now consider the sensitivity of the performance of the algorithm to other parameter choices and this will be illustrated using sample test problems. We will also study the effects of some of the modifications to the algorithm suggested by workers in the field. This will be achieved by considering the effects of these modifications individually on the behavior of the algorithm and comparing the results for some test problems. Specifically we will consider the importance of the choice of the parameters *a* and *b* on the performance of the algorithm and the alternatives methods for setting the values of the inertial weights, *w*. The first study compares the performance of the PSO algorithm using values of the parameters $a = 2.05$ and $b = 2.05$, called set 1, with the use of $a = 0.7$ and $b = 1.4$, called set 2. A swarm size of 30 is used with 800 iterations. The function considered is the Rastrigin function with 4 variables, denoted by RAS4. The results for 20 runs is given in Table 3.6. There is significant improvement with set 1.

As a graphical illustration, Figure 3.2 and Figure 3.3 show the progress of the iterations for an individual run for the two parameter sets. There is significant difference between the convergence paths but this could be because of the random nature of the process since it only involves one run of the PSO algorithm.

A further test considers how taking a specific value for the inertial weight compares with the continuous adjustment of the weight as the procedure continues. A swarm size

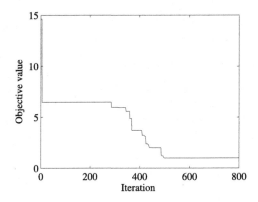

Figure 3.3 Changes in the objective function value for parameter set 2.

Table 3.7 Effect of constant and variable weight, w in finding the minimum of RAS4

Set	Mean	Best	Worst	St Dev
0.7	0.0497	0	0.9950	0.2225
Variable w	1.0452×10^{-7}	0	2.0905×10^{-6}	4.6744×10^{-7}

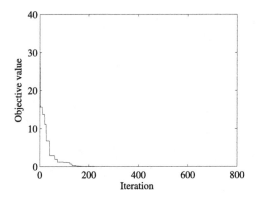

Figure 3.4 Iteration to determine the minimum of RAS4. $w = 0.7$.

of 30 is used with 800 iterations. The function considered is the Rastrigin function with 4 variables, RAS4. The results for 20 runs of the algorithm are given in Table 3.7. The table shows that there does seem to be a significant difference between the results, although further extensive testing should be performed before drawing hard and fast conclusions.

Figure 3.4 illustrates the progress of algorithm for the RAS4 function using w fixed at 0.7. Figure 3.5 shows the progress using w varied as the process proceeds.

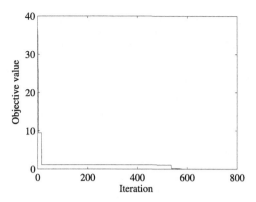

Figure 3.5 Iteration to determine the minimum of RAS4 using a variable w.

There is a difference between the nature of the graphs which may reflect the progress of the algorithm in the two cases. With the varying w the progress is initially rapid, then pauses for many iterations, until convergence is achieved, perhaps reflecting more searching for a better solution. Many more studies would be required to confirm this.

We now illustrate the performance of the swarm of particles graphically as they converge to the global minimum. This is shown by plotting the swarm particle locations on contour graphs of the Styblinski-Tang, Rastrigin and Rosenbrock functions in two variables. The position of the points at various stages of the iteration process is plotted on a series of contour graphs.

We begin with the Styblinski-Tang function by plotting the initial particle position values, then the position of the particles after fifty, one hundred and three hundred iterations, each set of particles on a separate contour plot of the function. This is shown in Figure 3.6.

We note that these graphs show rapid convergence to the global minimum of the function at the point $(-2.9035, -2.9035)$. In Figure 3.6, 3.7 and 3.8 the individual particles are shown by filled circles.

A second example illustrates the convergence behavior for the Rastrigin function which has many closely packed minima in the region $x = -5$ to 5, $y = -5$ to 5. Figure 3.7 shows the convergence of the swarm. For clarity only the locations of the many minima are shown, rather than a conventional contour plot. Effective convergence to the global minimum at $(0, 0)$ is achieved after three hundred iterations and convergence to the many nearby local minima is avoided.

As a final example we show how the algorithm behaves with Rosenbrock's function which has a single unique minimum at $(1, 1)$. The results for a run with the PSO algorithm for the optimization of Rosenbrock's function are shown in Figure 3.8. The contour graph of Rosenbrock's function shows a shallow banana shaped valley, it rep-

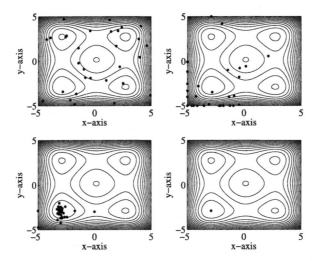

Figure 3.6 Contour plot of the Styblinski-Tang function in two variables, showing the progress of the particle swarm to convergence for 0, 50, 100 and 300 iterations.

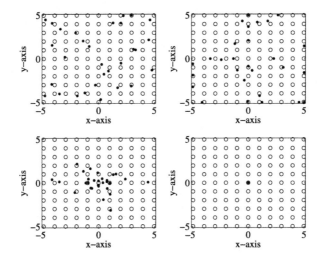

Figure 3.7 Rastrigin function in two variables, showing the progress of the particle swarm to convergence, for 0, 50, 100 and 300 iterations. The open circles show the many local minima.

resents a significant challenge for gradient methods because of its shallowness, however the PSO method copes well with this function. Although convergence is more gradual for this function, the points are converging on the required single minimum.

MATLAB scripts have been used throughout to implement these tests and to generate the graphical representation.

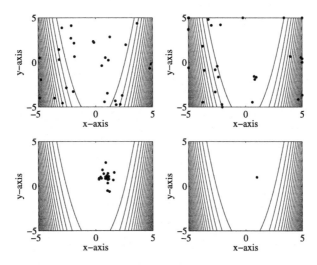

Figure 3.8 Contour plot of the Rosenbrock function in two variables, showing the progress of the particle swarm to convergence, for 0, 50, 100 and 300 iterations.

3.5 A REVIEW OF SOME RELEVANT DEVELOPMENTS

The method outlined in this chapter is one of the earlier biologically inspired algorithms to be used for optimization and has provided inspiration for the development of new algorithms. In this section we will consider recent developments in the use of the algorithm and describe the comparative studies of the particle swarm algorithm with other competitive methods.

It is important to test any new development on a standard set of test problems together with equitable setting of parameters so that the tests are fair. Tests need to be cleverly devised so that they check the key feature of the algorithm. That is to find global optima with good accuracy and reasonable efficiency. A key feature is to avoid becoming stuck at a local optima; a particularly challenging problem where local optima are tightly packed or the function profile provides rapid or near discontinuous change. Standard test problems reflect these features.

An interesting comparative study has been made between particle swarm optimization and GAs by Hassan et al. (2005). This study compared performance on some standard test problems, on a telescope array optimization and a satellite design problem and found both methods discovered good solutions for the problems but the particle swarming method was more efficient.

It is not possible to cover all the most recent developments in the field of PSO but we have selected papers that provide in our judgment interest and value and deserve further study. The following references outline some interesting developments:

We first consider the work of Arasomwan and Adewumi (2013). This work considers the problem of the best choice of inertial weights for the Particle Swarm Optimization algorithm. The use of inertial weights was an early development in the PSO algorithm which is generally accepted to provide an improvement to the basic algorithm. However over the years researchers proposed various alternative as to how these weights should be set and adjusted. The Arasomwan and Adewumi paper provides an assessment of these competing techniques, the authors point out it is important to test these competing methods in a completely equitable way, a difficult aim to achieve. They review and compare a range of methods which we will briefly describe.

Rather than using a fixed weight an early development was proposed by Eberhart and Shi (2000) which suggested a simple linear adjustment to the weight which reduced its value as the algorithm proceeded: called Linear Decreasing Inertia Weight (LDIW-PSO) procedure. This modification has been described earlier in this chapter. However it is found with this modification that the algorithm may still get stuck at a local optimal point rather than find a global optimum. Consequently a simple adjustment was suggested to the LDIW method. In this method introduced by Feng et al. (2007) a further chaotic factor is introduced that modifies the end weight value. The authors of the method called the Chaotic Descending Inertial Weight PSO (CDIW-PSO) reported significant improvement in the algorithms performance. Another method suggested by Gao and Duan (2007) called the Random Inertia Weight and Evolutionary Strategy PSO, (REPSO), introduced an interesting combination of PSO and Simulated Annealing. The weights being chosen at each stage using an annealing probability. Again Gao and Duan reported good results for this algorithm. Optimization problems of high dimension and a multi peak profile present challenging optimization problems. The method of Dynamic Particle Swarm Optimization, (DAPSO) introduced by Xin et al. (2009) was introduced to deal with premature convergence and provide good convergence speed for this type of problem. It modified the standard linear weight adjustment formula using factors using estimates of particle group fitness at each iteration. Promising results were reported for this method. Another method considered in this paper was adaptive swarm optimization introduced by Alfi and Modares (2011). In this algorithm the weights were adjusted at each iteration using a function of the fitness of the current best solution. In addition, particles are chosen for mutation and a Gaussian random value added to the particle. The final two variants considered by Arasomwan et al. were Dynamic Nonlinear and Dynamic Logistic Chaotic Map PSO, (DLPSO2). These were introduced by Liu et al. (2009). One method introduced a dynamic nonlinear factor to the standard linear weight adjustment formula. In addition it used a variant of the chaotic map technique given in the CDIW-PSO method, providing a multi-faceted search procedure. Liu et al. reported good results for this method. Clearly an independent analysis of these competing claims would be useful.

Arasomwan et al. embarked on a carefully designed and detailed comparative study on the methods we have briefly described above. The authors maintain that to provide a fair comparison for the LDIW method its parameters must be properly set. With this setting of the LDIW parameters the authors provide a range of numerical results for the comparison of the methods described with the LDIW method for range of test problems. Their conclusion is that the linear dynamic inertia weight adjustment introduced by Shi and Eberhart (1998) with appropriate parameter settings performs competitively with all the variants described above.

The reader is advised to make their own judgments about algorithm performance by referring to the papers we have cited. The paper by Arasomwan et al. does highlight the difficulties of comparative studies and the different conclusions that may be reached.

Another paper we discuss addresses the major problem of global optimization algorithms, that is how to avoid premature convergence to a local optimum. The work of Napoles et al. (2012) addresses the fundamental problem of global optimization methods which is premature convergence to a global optimum, from which the algorithm is unable to escape. This is frequently possible since in many problems the locations of the optima may be tightly packed or distantly isolated from the other minima. Two questions arise when dealing with this problem: these are:

1. How can the fact that the algorithm is trapped near a local optimum be detected?
2. How can the algorithm be stimulated to move to search a wider area of territory in which the function is defined?

A simple approach to the first question is to declare that the algorithm is trapped, if after a large number of iterations, there is no improvement in the value of the function value, but this could mean the optimum has been reached and it is unclear how to set the specific value of the large number of iterations. Since this approach has obvious shortcomings a more subtle approach is needed. One method involves the use of the criteria for judging if premature convergence has occurred. This is by using the maximum radius of the swarm. Figure 3.6 illustrates how the swarm gradually clusters around the optimum point. Clearly this may reduce the diversity of the search process and it is important to be able to detect this state. To do this, according to Napoles et al., we define the point $\mathbf{x}_g$ as the global best point in the neighborhood and $\mathbf{x}_i^{(k)}$ as the position of the other particles for $i = 1, , ..., n_{pop}$ at iteration k in the swarm then the expression:

$$\| \mathbf{x}_i^{(k)} - \mathbf{x}_g \| \text{ for } i = 1, 2, ..., n_{pop} \tag{3.14}$$

gives the Euclidean distance between the global best point and each of the particles in the swarm. Here n_{pop} is the population size. Consequently the expression:

$$\max_{i=1:n_{pop}} \| \mathbf{x}_i^{(k)} - \mathbf{x}_g \| \tag{3.15}$$

represents the maximum distance of a typical particle in the swarm from the global best particle found. Let r_{max} and r_{min} be the range of possible values for particle position values, then the we can normalize the maximum distance as follows:

$$s^{(k)} = \frac{\max_{i=1:n_{pop}} \| \mathbf{x}_i^{(k)} - \mathbf{x}_g \|}{|r_{max} - r_{min}|} \tag{3.16}$$

Thus if the value of $s^{(k)}$, the current radius of the swarm, is less than a preset value, premature convergence is deemed to have occurred since the points are relatively closely clustered around the current best point. This is used to distinguish a premature convergence state from a global one.

Napoles et al. (2012) proposed to deal with the second question of moving to a better optimum by a process he calls random sampling in variable neighborhoods. The concept of separate neighborhoods was introduced by Hansen and Mladenovic (2001). The aim of this is to allow the particles to move so that a better point may be found giving an improved objective function value. However the movement must be conducted in an organized way, so that good information is not lost and that the process is not trapped again.

Napoles et al. defined the ranges for the new sub-regions and selects uniformly distributed random values for each dimension of the points in each of these new ranges. Taking the union of these points or suitable subsets of these points produces the new set of points for the swarm. The selected subsets are described by the Napoles et al. as points selected as good enough particles by an elitist criterion. The reader is referred to Napoles et al. for details of this algorithm.

Napoles et al. also discussed a simple variant of the algorithm we have described which instead of using the radius of the swarm uses preset maximum number of function evaluations to determine if the algorithm is trapped. This is a less costly procedure in terms of computation time and they report it to be fully competitive with the form of the algorithm using the swarm radius. Further tests were performed by the author to compare the performance of the algorithm with other variants of the PSO algorithm, the attraction-repulsion based PSO, the Quadratic Interpolation based PSO, the Gaussian Mutation based PSO and a hybrid variant of the PSO using simulated annealing. Napoles et al. state that the new algorithm produces superior results in most cases.

3.6 SOME APPLICATIONS OF PARTICLE SWARM OPTIMIZATION

Particle Swarm Optimization (PSO) is a well established algorithm and is often cited in the literature and reported to have been applied to solve efficiently numerous problems which arise in real life. Here we indicate the nature of a small selection of these. In papers by Kang et al. (2012) and Gökdağ and Yildiz (2012) beams are modeled using the finite element method and damage is expressed by stiffness reduction at a global level. It is reported that the resultant optimization problem has been successful solved using the PSO algorithm.

In a paper by Hassan et al. (2005) the authors describe an application to minimize the weight of a simple gearbox for a light aircraft. In addition, they describe the optimum design of the distribution of a group of small telescopes so that they jointly provide the resolution of a larger, single telescope. This is an important problem for astronomers dealing with scarce financial resources and requiring optimum observational results.

In his MSc thesis, Talukder (2011) has produced a useful listing of applications of the PSO method, include many Biomedical applications; for example the diagnosis of Parkinson's through body tremors. He also described applications in robotics, including the development of the robot running process, and to problems in graphics, including character recognition.

3.7 SUMMARY

The method outlined here is one of the earlier biologically inspired algorithms to be used for optimization of non-linear functions. It has provided a useful impetus in the development of new algorithms inspired from many other facets of the manner in which the natural world agents seek to solve problems efficiently.

We have described the nature of the basic PSO algorithm and some of the many suggested changes to this algorithm to improve its performance. To examine the efficiency of the basic PSO algorithm we have tested its performance on a number of standard test problems, some of which have multiple minima, and have examined how effective the method is in finding the best of these minima, the global minima.

In addition the some of the modifications we have described have been tested systematically on the same standard test problems to detect if any significant improvements have been achieved. It is always dangerous to draw hard and fast conclusions from such tests but they give some insight into the nature of the algorithm and the effects of specific modifications. These studies are illustrations rather than pieces of research.

3.8 PROBLEMS

3.1 We wish to minimize the objective function $f(\mathbf{x}) = x_1^2 + x_2^2$. Assuming an initial particle swarm population $\mathbf{x}_1 = [2.5, \ 1]^\top$ and $\mathbf{x}_2 = [2, \ 1]^\top$ calculate the objective function values for this population and hence $\mathbf{x}_{lopt}$ and $\mathbf{x}_{gopt}$.

3.2 Use the same objective function and initial population as in Problem 3.1. Take the initial velocity vector as $[0, \ 0]^\top$ and set $a = 1$, $b = 1$. Use the random vectors $\mathbf{r}_1 = [0.23 \ 0.45]^\top$ and $\mathbf{r}_2 = [0.42 \ 0.65]^\top$ and perform one iteration of the PSO algorithm using (3.2) and (3.3). Note: After the value of $\mathbf{x}$ has been updated you must update $\mathbf{x}_{lopt}^{(0)}$ and $\mathbf{x}_{gopt}^{(0)}$.

3.3   Taking $w^{(0)} = 1$ and $w^{(t_{max})} = 0.2$ where $t_{max} = 10$ use equation (3.5) to generate a range of weights between the values 1 and 0.2.

3.4   Plot a graph showing how v_{max} varies with iteration t using formula (3.11). Take $t_{max} = 10$ and hence $t = 0 : 10$. Draw three graphs taking: $p = 1, 2, 3$.

CHAPTER 4

The Cuckoo Search Algorithm

4.1 INTRODUCTION

The Cuckoo search algorithm is based on the parasitic nature of the cuckoo which lays its eggs in other birds' nests thus releasing itself from the burden of direct parenthood and utilizing the efforts of the host bird for its own gain. The cuckoo wishes to optimize the likelihood that its offspring survives and prospers and adopts various strategies to achieve this outcome. For example, the cuckoo camouflages the egg so that it does not stand out from others in the nest and places the egg at a time so that it hatches first, thereby gaining a survival advantage. However, there is the possibility that the egg will be discovered and removed from the nest or the nest will be abandoned and another built in a different location to replace it.

The cuckoo search algorithm was introduced by Yang and Deb (2009). They used the above concepts but refined them to create an optimization algorithm based on the cuckoos behavior.

The key rules of this algorithm are:

1. Each cuckoo lays one egg at a time, and places its egg in a randomly chosen nest.
2. The best nests with high quality eggs will carry over to the next generation, the fitness or quality of the eggs being determined by the objective function.
3. The number of available host nests is fixed, and the egg laid by a cuckoo is discovered by the host bird with a specific probability. Then the egg is either thrown away or the nest is abandoned a new nest built to replace it.

Each of the cuckoo eggs in a nest at a specified location is a possible solution and each new additional placed cuckoo egg represents a new possibly better solution. The success of a particular egg placement can be judged using a specified fitness or objective function. Yang and Deb point out that for implementation purposes since only one egg is placed in a nest in practice there is no distinction between nests and eggs, both having the same location.

An additional important feature introduced by Yang and Deb was to generate new egg/nest locations using a random walk based on Lévy flights. The intention is that using Lévy flights better represents the natural search process for the new nest locations for egg placement. Lévy flights are generated from a specific distribution which allows large exploratory steps to be taken as the algorithm proceeds in the hope of ensuring a thorough exploration of the search space. Since small steps may also occur the balance

Introduction to Nature-Inspired Optimization
DOI: 10.1016/B978-0-12-803636-5.00004-9
69

between exploratory steps and smaller steps when the algorithm is homing in on the global minimum can be maintained. Lévy flights have been reported to better reflect the exploratory nature of certain natural insect flights.

The Lévy distribution is described in more detail in Chapter 1. We can now describe formally the nature of the Yang and Deb algorithm in the following section. This is based on the more recent paper by Yang and Deb (2014).

4.2 DESCRIPTION OF THE CUCKOO SEARCH ALGORITHM

Using the rules 1, 2 and 3 of Section 4.1 described above an algorithm can be devised as follows:

Step 1: Generate a population of n_{pop} of nest for eggs at $\mathbf{x}_i$ where $i = 1, 2, ..., n_{pop}$

Step 2: For a preset number of generations or iterations: generate a new set of nest locations based on a random walk procedure where each random step length is selected from the Lévy distribution for each nest i. The new $\mathbf{x}_i$ values are obtained from:

$$\mathbf{x}_i^{(t+1)} = \mathbf{x}_i^{(t)} + \alpha \mathbf{r}_l \tag{4.1}$$

where t is the generation, the value $\mathbf{r}_l$ is a vector of random numbers, selected from a Lévy distribution and α is a scaling factor.

Then evaluate the fitness of the new $\mathbf{x}_i$ nest location using the fitness or objective function $f(\mathbf{x}_i)$.

Select randomly a new nest $\mathbf{x}_j$. If $f(\mathbf{x}_i) > f(\mathbf{x}_j)$, then assuming we are minimizing the fitness function replace $\mathbf{x}_i$ by $\mathbf{x}_j$, since this provides an improvement in the fitness function.

Step 3: A randomly selected proportion of the worst nests are abandoned, determined by a user set constant P in the range 0 to 1. This random replacement is implemented by the equation:

$$\mathbf{x}_i^{(t+1)} = \mathbf{x}_i^{(t)} + \alpha s(\mathbf{x}_j^{(t)} - \mathbf{x}_k^{(t)}) \tag{4.2}$$

Parameter s is the step size and Yang and Deb suggest a value of one in their 2009 paper for this parameter but alternatives have been suggested and the parameter may be problem dependent.

Here $\mathbf{x}_j^{(t)}$ and $\mathbf{x}_k^{(t)}$ are randomly selected locations. The best nests are determined according to their values assessed by the fitness criteria.

Step 4: Rank the solutions according to fitness and find the current best solution.

Step 5: Repeat from Step 2 until the stopping criterion is met or number of generations is completed.

It is encouraging that the algorithm has relatively few parameters and its structure is relatively simple. Although the Yang and Deb suggest that the value of s may be taken as one, other authors suggest smaller values to avoid the algorithm wandering outside the search region. The decision on the value of α is likely to be problem dependent and the choice may be related to the scale of the specific problem. One way of doing this is to set α to values of order $L/10$ or $L/100$, where L is the scale of the problem, see Yang and Deb (2014).

This is the basis of the algorithm and Yang and Deb have reported good results for tests of the cuckoo search algorithm's effectiveness in locating the global minima on a demanding range of standard non-linear minimization test problems. The algorithm has been applied to many real-world applications.

In practice, the implementation of the Lévy distribution may be achieved using the algorithm of Mantegna (1994). It should be noted that the Mantegna algorithm requires significant computational effort. See the paper by Leccardi (2005) for alternative methods of implementing the Lévy algorithm and an examination of their efficiency.

4.3 MODIFICATIONS OF THE CUCKOO SEARCH ALGORITHM

An interesting modification of the original Cuckoo search algorithm has been introduced and tested by Tawfik et al. (2013). We now describe this modified algorithm. Tawfik et al. call their algorithm the one rank cuckoo search algorithm and it involves two amendments to the basic algorithm. The first amendment replaces the two evaluations of fitness of the original algorithm which occur before and after the replacement of a proportion, P, of the nests, by a single re-evaluation and ranking in order of fitness, which occurs after the generation of new nests using Lévy flights and replacement. This reduces the number of function evaluations by merging the exploration and exploitation phases. A new parameter r_{or} is introduced that is initially set at 1 and gradually reduced using the equation:

$$r_{or}^{(t+1)} = r_{or}^{(t)}(1 - 0.5/n_{var}) \tag{4.3}$$

where t is the generation or iteration number and n_{var} the number of variables in the problem. Thus the larger the number of variables the more gradual the change in r_{or}. This reflects the fact that in general the difficulty of the problem increases with the number of variables. A reduction is implemented if there has been no improvement in the solution for t_{or} generations where t_{or} is set by the user.

The second amendment notes that due to the use of Lévy flights, steps may be allowed that take the solution beyond the region in which the problem has been defined. To avoid this problem the Tawfik et al. introduce a parameter r_b where r_b is defined by the equation:

$$r_b = 1 - 1/\sqrt{n_{var}} \tag{4.4}$$

If a new solution is generated which is outside the required bounds, then if the value r_b is greater than a uniform random value, the new solution is replaced by a randomly selected solution from the current best. Otherwise the new solution is replaced by a random point satisfying the constraints of the problem. That is one lying within the region in which the problem is defined. Tawfik et al. report improved results using this amended algorithm for a range of test problems. **Algorithm 2** provides an outline implementation.

The next modification to the cuckoo search algorithm has been introduced by Fateen and Bonilla-Petriciolet (2014). They call this the gradient based Cuckoo search for global optimization.

These authors note that it is an advantage of the cuckoo search algorithm and other similar algorithms that they do not require the evaluation of the gradient by users. However Fateen and Bonilla-Petriciolet propose that if the gradient is readily available it may improve the efficiency of the basic cuckoo algorithm by taking account of the information the gradient supplies. The authors reported the results of numerical studies they carried out which show significant improvements in performance on some standard test problems.

The modification they suggest is not difficult to implement and we now describe it. The step at which some nests are abandoned involves their replacement using the equation:

$$\mathbf{x}_i^{(t+1)} = \mathbf{x}_i^{(t)} + \mathbf{U} \circ (\mathbf{x}_j^{(t)} - \mathbf{x}_k^{(t)}) \tag{4.5}$$

Setting:

$$\lambda = \mathbf{U} \circ (\mathbf{x}_j^{(t)} - \mathbf{x}_k^{(t)})$$

and denoting the gradient of the function by $\mathbf{g}$ in the usual notation we have

$$\mathbf{g} = \nabla f(\mathbf{x}_i)$$

Fateen and Bonilla-Petriciolet propose (4.5) is replaced by:

$$\mathbf{x}_i^{(t+1)} = \mathbf{x}_i^{(t)} + \lambda \times \text{sign}(-\lambda/\mathbf{g})$$

The last term of this equation means that we take the sign of the negative ratio of the step divided by the gradient of the objective function. In this way, Fateen and Bonilla-Petriciolet state, the direction of the step is taken towards the minimum. This modification is a simple one to implement and does not involve additional parameters. The gradient must be supplied by the user and does not change the stochastic nature

---

Algorithm 2 Cuckoo Search with Tawfik et al. modification.

Initialize host nests at random locations

Set the initial parameters t, r_{or} and t_{or} all equal to 1

while $t < t_{max}$ **do**

    for $i = 1$ to n_{pop} **do**

        $\mathbf{x}_i^{(t+1)} = \mathbf{x}_i^{(t)} + s\mathbf{N}_i \circ \mathbf{l}_i \circ (\mathbf{x}_i - \mathbf{x}best)$

    end for

    x_{best} is the current best value

    if $u < r_{or}$ **then**

        for $i = 1$ to n_{pop} **do**

            if $u < P$ **then**

                abandon nest and replace with a new solution at a new location

                $\mathbf{x}_i^{(t+1)} = \mathbf{x}_i^{(t)} + \mathbf{U} \circ (\mathbf{x}_j^{(t)} - \mathbf{x}_k^{(t)})$

                Where $\mathbf{U}$ is a uniformly generated vector of random numbers and $\mathbf{x}_j^{(t)}$, $\mathbf{x}_k^{(t)}$ are randomly selected solutions

            end if

        end for

        Depending on the value of r_b calculated from (4.4) the new solution is either bounded by the best solution or it is ensured that it lies within the defined region. Evaluate the fitness of the new solutions by using the objective function. Rank all the nests and find the best nest and its fitness value and store the values.

    else

        Bound points within defined region. Evaluate new nests and rank them keeping the best. Replace fraction and repeat evaluation.

    end if

    Get current best solution. If no improvement for t_{or} generations decrease r_{or} using (4.3) if stopping criteria not met continue iterations.

end while

Finish and output best result.

---

of the algorithm although there is a significant demand on the user as the derivatives must be given. We give the symbolic evaluation of the gradient for Rastrigin's function. Rastrigin's function is defined by the equation:

$$f(\mathbf{x}) = An_{var} + \Sigma_1^{n_{var}}[x_i^2 - A\cos(2\pi x_i)]$$

The components of the gradient of this function are given by

$$\partial f_i / \partial x_i = 2x_i - 2\pi A \sin(2\pi x_i) \quad \text{for } i = 1, 2, ..., n_{var}$$

Thus

$$\mathbf{g} = \left\{ \begin{array}{l} 2x_1 - 2\pi A \sin(2\pi x_1) \\ 2x_2 - 2\pi A \sin(2\pi x_2) \\ \cdots\cdots \\ 2x_n - 2\pi A \sin(2\pi x_{n_{var}}) \end{array} \right\}$$

The gradient is symbolically evaluated by the user and included as a function within the program. A numerical approximation of the gradient could also be considered.

A further modification of the basic cuckoo search algorithm has been suggested by Valian et al. (2011) who propose that key parameters of the algorithm be adjusted from generation to generation thereby taking account of the progress of the algorithm, hence providing improved tuning of the parameters as the process proceeds. The parameters of the cuckoo search algorithm are essentially the values of P, the proportion of abandoned nests, and α, the step size scaling factor. These are the parameters for which they provide an adjustment procedure.

For their modification they introduce a maximum and minimum value for the parameters P and α as:

$$P_{min} \leq P \leq P_{max} \text{ and } \alpha_{min} \leq \alpha \leq \alpha_{max}$$

Using these bounds they suggest the following procedure for adjusting the values of the parameters P and α at iteration t:

$$P^{(t)} = P_{max} - \frac{t}{t_{max}}(P_{max} - P_{min}) \tag{4.6}$$

Here $P^{(t)}$ is the value of P at iteration t and t_{max} the maximum number of generations.

The effect of this equation is that for $t = 0$, P takes its maximum value and when the iteration, $t = t_{max}$, P takes its minimum value. The value of P decreases linearly between these two values as t increases. The adjustment of α is achieved using the equation:

$$\alpha_t = \alpha_{max} e^{ct} \tag{4.7}$$

where α_t is the value of α at the current iteration t and the c is defined by the equation:

$$c = \log_e \left(\frac{\alpha_{min}}{\alpha_{max}} \right) / t_{max} \tag{4.8}$$

Since the value of c is negative the value of α_t will decrease as t increases. In this way the authors indicate the values of P and α can start at large values allowing a full exploration of the search space and then be decreased as the process continues. The graph in Figure 4.1 shows the manner in which α changes using this method.

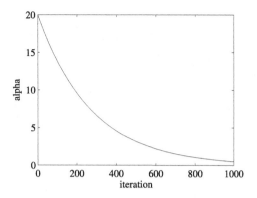

Figure 4.1 Shows the change in the value of α as the number of iterations increases. See (4.7) and (4.8).

Figure 4.1 shows that α changes from the maximum value 20 to the minimum value 0.5 in an exponentially decreasing fashion.

Valian et al. report improvements in results when applied to two bench mark training sets for neural network problems when compared with the original cuckoo search algorithm. This modification would benefit from further studies using the standard global optimization test problems. There are many other modifications that have been suggested and those which we have described are a selection of these that provide further opportunity for study and development.

4.4 NUMERICAL STUDIES OF THE CUCKOO SEARCH ALGORITHM

We now study the nature of the cuckoo search algorithm by carrying out numerical studies of some of the key features of the algorithm. These studies are not meant to be exhaustive but we hope to provide some insights into the performance of the algorithm. They are certainly not as comprehensive as would be expected in a specific research paper.

The first study examines how the number of nested eggs affects the performance of the algorithm in optimizing the Rastrigin function with 8 variables. The study is performed for 10, 20, 30 and 40 nested eggs; a low number of 600 iterations is used since it is the comparative performance we wish to examine. Apart from the number of nested eggs the same set of parameters. $P = 0.25$, $s = 0.1$, are used in each run. Table 4.1 gives the results obtained using 20 runs of the algorithm.

Since we know the minimum of Rastrigin's function is zero, this table shows a gradual improvement in the results obtained as the number of nested eggs increases. The difference between the results for 10 nested eggs and 40 nested eggs is quite marked.

Table 4.1 Showing the effects of the number of nests on the convergence of the CSA minimizing the eight variable Rastrigin function, RAS8

Nests	Average	Best	Worst	St Dev
10	0.4637	7.7150×10^{-5}	1.2240	0.5018
20	0.4215	5.4288×10^{-5}	1.7412	0.5515
30	0.0258	6.2416×10^{-5}	0.1863	0.0490
40	0.0088	1.9524×10^{-5}	0.0873	0.0198

Table 4.2 Minimization of Rastrigin's function with 8 variables, (RAS8) showing the effect of the number of iterations on the accuracy obtained

Generations	Average	Best	Worst	St Dev
200	4.7890	3.2949	6.8610	0.9417
500	0.4002	1.9689×10^{-4}	2.0802	0.5724
1000	3.6783×10^{-7}	0	7.3560×10^{-6}	1.6449×10^{-6}

Further comprehensive studies would need to be performed to provide added validity to these conclusions.

We indicate above that we had performed the tests given in the table using a fairly low number of iterations or generations in the next study we consider the effect of the number of iterations on the accuracy of the solution, these results are given in Table 4.2. For this study we use 30 nests and apart from the number of generations we use the same set of parameters and minimize Rastrigin's function with 8 variables. Twenty runs are performed for each number of iterations.

Table 4.2 shows that the number of iterations used has a very significant effect on the accuracy of the solutions, there is an improvement with every increase in the number of generations. In particular the last row of the table which gives the results for a large increase in the number of generations shows very good results.

However the effectiveness of many algorithms is problem dependent and also depends on the number of dimensions or variables which the function involves and extremely high number of iterations may be required for more formidable problems with many variables. The next study will consider the effect of dimensionality on the accuracy of the results.

In Table 4.3 we compare the results of minimizing the Rastrigin function for various numbers of variables. One thousand iterations are used and thirty nested eggs and the same set of parameters for each set of twenty runs. RAS6, RAS8, RAS10 and RAS15 represent Rastrigin's function with six, eight, ten and fifteen variables respectively. We use 20 runs of the algorithm. Table 4.3 shows that the accuracy of the results tail off very rapidly with increasing dimensionality. Improved results can be obtained by using larger numbers of iterations. Note the poor result obtained for the problem with fifteen variables.

Table 4.3 This shows the effect of increasing dimensionality on the accuracy with which solutions can be obtained

Problem	Average	Best	Worst	St Dev
RAS6	0	0	0	0
RAS8	1.0970×10^{-10}	5.6843×10^{-14}	8.8639×10^{-10}	2.0942×10^{-10}
RAS10	0.0038	3.2870×10^{-6}	0.0245	0.0077
RAS15	8.9229	5.5809	13.4243	1.9411

Table 4.4 Showing the effect of the proportion of nests abandoned, P on the accuracy of the global minimum obtained for the eight variable Rastrigin function, RAS8

P	Average	Best	Worst	St Dev
0.10	0.2564	2.8271×10^{-4}	1.0572	0.3958
0.25	0.0354	1.5307×10^{-6}	0.4172	0.0947
0.50	1.6093	0.1620	2.9563	0.7588
0.75	3.4078	1.8550	5.6592	0.9488

Table 4.5 Minimization of the function RAS15

P	Mean	Best	Worst	Standard Deviation
0.10	2.7685×10^{-4}	5.5825×10^{-9}	0.0020	6.0428×10^{-4}
0.25	1.5149	0.0016	3.8319	1.4054

The cuckoo search algorithm has few parameters to set. However one parameter that may be critical is the proportion of nests to be abandoned, denoted by P.

The following study shows the effect of different values for the parameter P on the accuracy of the results and is shown in Table 4.4. Six hundred iterations are used together with 30 nests and the test function we minimize is Rastrigin with eight variables. The results are given for values of the parameter $P = 0.1$, 0.25, 0.5 and 0.75 and twenty runs of the algorithm are performed.

Table 4.4 shows the value $P = 0.25$ which produces the best results and is often recommended; the other values are relatively close together so few sound conclusions can be drawn except that further investigation is warranted using a wider range of problems and more algorithm runs. It must also be noted that the effects of many parameters are problem dependent, i.e. some parameter choices are good for some problems but not for others. It must be noted the effects of the value of P may be problem dependent. Hence for larger numbers of variables or different types of problem another value of P may be better. Table 4.5 illustrates this.

Table 4.5 shows the results of minimizing Rastrigin's function with 15 variables for the value setting the proportion of abandoned nests at $P = 0.1$ and $P = 0.25$. It

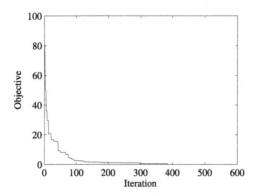

Figure 4.2 The optimization of the Rastrigin function with six variables. Plot of objective function value against iteration showing convergence of to the solution.

uses $s = 0.1$ and 2000 iterations or generations. The results are for twenty runs of the algorithm.

Clearly for the Rastrigin function with 15 variables the value $P = 0.1$ gives a major improvement in the results obtained using $P = 0.25$. For low numbers of variables this difference is not pronounced.

We now give some graphical illustrations of the progress of the algorithm in solving global minimization problems. Note that these graphs and results were obtained using step size of $s = 0.1$, $P = 0.25$, six hundred iterations and numbers of nested eggs was thirty.

Figure 4.2 shows the convergence of the cuckoo search algorithm to the global minimum of the six variable Rastrigin's function which is a formidable problem because of the dimensions of the search space and the number of local optima. Although little can be gleaned from one graph it does appear to show blocks of iterations where no change is occurring in the objective but ultimately followed by a significant drop in the function value. This may imply an element of searching the surrounding region before finding an improved objective function value.

In comparison, we now show a graph of the convergence of the algorithm for the two variable Rosenbrock function given in Figure 4.3 using only 600 iterations.

Figure 4.3 shows how the search rapidly finds the minimum of Rosenbrock function. Since this function has only one minimum, the figure displays a relatively continuous reduction in the function value.

The next pair of figures show the effect, if any, of using different parameter values for the proportion of abandoned nests P, for solving the eight variable Rastrigin function (RAS8). The values P used are 0.1 and 0.25. The results are shown in Figures 4.4 and 4.5. Figure 4.4 with $P = 0.25$ and Figure 4.5 with $P = 0.1$.

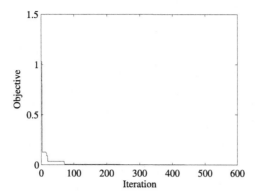

Figure 4.3 The optimization of the 2 variable Rosenbrock function. Plot of objective function value against iteration, showing the convergence to the solution zero.

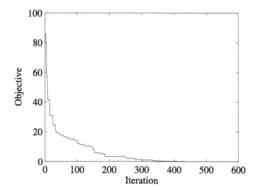

Figure 4.4 Plot of convergence of Cuckoo search algorithm for optimization of Rastrigin's function with eight variables using $P = 0.25$.

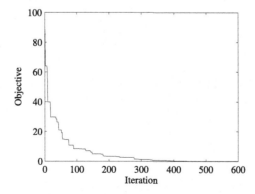

Figure 4.5 Plot of convergence of Cuckoo search algorithm for optimization of Rastrigin's function with eight variables using $P = 0.1$.

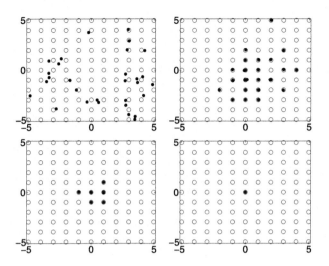

Figure 4.6 Plot of population convergence of Cuckoo search algorithm for optimization of Rastrigin's function with two variables. The subplots show the locations of points at the initial step then at the 50th 200th and 800th iteration where convergence of the algorithm has occurred.

Comparing Figures 4.4 and 4.5 shows there is little discernible difference in performance of the algorithm that is shown in the two figures.

Figures 4.6 and 4.7 show how the eggs in the locations of the selected nests change as the algorithm proceeds; a process similar to the changes in particle swarm locations as they converge on the solution.

Figure 4.6 shows the process in terms of egg locations for the minimization of Rastrigin's function which has many local minima. Similarly, Figure 4.7 shows the convergence of the algorithm for the Styblinski-Tang function which has four well defined local minimum.

The global minimum for Rastrigin's function lies at the point [0, 0], with objective function zero. The contour graph shows how that after some initial exploration of the region in particular some of nearby local minima the algorithm converges to global minimum at the point [0, 0].

The global minimum for the Styblinski-Tang function lies at [−2.9035, −2.9035], with objective function −78.3223. This function has four clearly defined well separated minimums and does not present a major challenge for the cuckoo search algorithm but does illustrate that the global minimum is found.

The contour graph shows that after some initial exploration of the region in particular some of nearby local minima the algorithm rapidly converges to global minimum at [−2.9035, −2.9035]. In the case of the Styblinski-Tang function this has been achieved using relatively few iterations and the algorithm has converged by the 200th iteration.

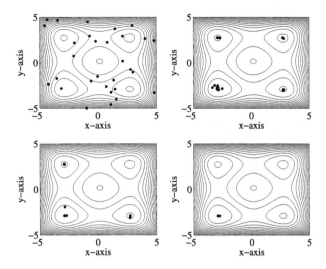

Figure 4.7 Contour plot of the of Styblinski-Tang function with two variables. Plot shows the convergence of the population of Cuckoo search algorithm to the minimum of the function. The subplots show the locations of nests at the initial step then at the 50th, 100th and 200th iteration.

We now consider some recent developments of the cuckoo search algorithm which provide applications beyond the basic non-linear optimization problem.

4.5 EXTENSIONS AND DEVELOPMENTS OF THE CUCKOO SEARCH ALGORITHM

This first development of the basic algorithm we consider has been suggested by Erik Cuevas and Reyna-Orta (2014). They call their algorithm multi-modal Cuckoo search. We will mention this only briefly since it goes beyond the specific problem of finding the global minimum of a function. It does show how to develop the cuckoo search algorithm to find not only the global minimum but also to register and store the local minima that are located in the search process and which may be of interest in a full analysis of the multi-modal behavior of the objective function; thus providing deeper insights into the nature of the function being optimized.

Another application shows how cuckoo search can be used to solve the Traveling Salesman Problem. Introduced by Quaraab et al. (2014). They describe their algorithm as a Discrete Cuckoo search algorithm for the Traveling Salesman Problem (TSP). The TSP is simply described as finding the route giving the minimum distance for a sales person to travel, visiting a number of towns only once and return to the starting point. Although simple to describe, solving the problem is very difficult and the computational difficulty of the problem increases rapidly with the number of towns that must be visited.

In fact the problem is formally classified as NP hard which places it amongst the most difficult of mathematical problems.

There are two aspects to the paper of Quaraab et al. They describe how to apply the cuckoo search algorithm to solve the TSP problem but they also introduce modifications to the basic algorithm to improve its performance. The improvement is that the cuckoo population is divided into three types:

1. Cuckoos starting from the current best position and who are seeking new areas with improved solutions do so much better than cuckoos randomly generated.
2. A proportion of the Cuckoos that are exploring the region far from the current best solution.
3. A proportion of the Cuckoos that are exploring the region around the current best solution and trying to improve it.

Quaraab et al. characterize these as zones and propose that this will allow good local searches but avoid being stuck in a local minimum since the algorithm moves from zone to zone using Lévy flights so that a thorough exploration of the region will be achieved.

Quaraab et al. (2014) provide a very extensive investigation into the methods efficiency by solving a very varied set of standard TSP test problems and report significantly improved results when compared with other approaches to solving the TSP problem.

4.6 SOME APPLICATIONS OF THE CUCKOO SEARCH ALGORITHM

In a paper by Valian et al. (2011), the authors describe how the CSA has been applied to the efficient implementation of neural network for solving problems related to the analysis of data sets arising from breast cancer diagnosis. Determining the weights in a Neural Networks (NN) requires global optimization.

Mohamad et al. (2014) have applied the CSA to improve the efficiency of the application of Neural Networks to estimating market clearing prices. In addition, a number of other applications are cited these include: object oriented software testing, pattern recognition, optimum job scheduling and data fusion for wireless sensors networks.

Dash and Mohenty (2014) describe how the Cuckoo Search Algorithm can be used for speech recognition by matching based on extracted optimum features of speech. We conclude this review of applications by considering the paper by Taherian et al. (2013). Here Taherian et al. use the Cuckoo Search to efficiently implement the neural networks applied to electricity price forecasting clearing prices. The global nature of the CSA search for the optimal solution provides promising results.

4.7 SUMMARY

In this chapter we have described the Cuckoo search algorithm and how it is based on specific features of the cuckoo's nesting behavior. In particular we have described how the Lévy distribution plays a major role in the implementation of the algorithm. In addition, we have discussed certain modifications of the original cuckoo search algorithm which have been suggested to improve its performance. The results of small scale numerical studies have been given which study the role of key parameters on the behavior of the algorithm. Finally some graphical illustrations have been provided to show how the algorithm converges to the optimum of specific test functions.

4.8 PROBLEMS

4.1 For the cuckoo search algorithm, assume an initial population of nests with eggs at locations: [1, 1], [−1, 2], [0, 1] and [1.5, 2]. Generate a new population of nests using (4.1). You may assume $\alpha = 0.1$ and that the following values have been generated from the Lévy distribution: [0.15, −0.5], [1.25, 0.25], [1.5, 2] and [−0.5, 0.5] corresponding to each member of the population $i = 1$, 2, 3 and 4. The objective function that we wish to minimize has the form $f(\mathbf{x}) = x_1^2 + x_2^2$. For each of the four new points you have generated find the value of this objective function and rank the values. Hence find the best estimate of the minimum of $f(\mathbf{x})$ for this generation. An elitist strategy selects the best result for each nested egg by comparing the original egg locations and new egg locations using the objective function value; if an improvement is obtained the new point replaces the original point otherwise the original point is retained. Use this strategy to generate the new generation of egg locations. Note you will have to calculate the objective function values for the original population.

4.2 Using the results you have obtained for Problem 4.1 and selecting two nests arbitrarily to be abandoned, for these nests calculate new locations using (4.5). Note that this is an alternative to the original algorithm. For the random numbers you require select them arbitrarily in the range [0, 1]. In addition select the points j and k arbitrarily from the list [1, 2, 3, 4]. For the new points you have generated find the value of the objective function $f(\mathbf{x})$ defined in Problem 4.1. Using the objective function find the best value of the objective function generated so far.

4.3 Find the gradient of the following functions $f(\mathbf{x})$ at the point [0.5, 0.5]: (i) $f(\mathbf{x}) = x_1^2 + x_2^2$, (ii) $f(\mathbf{x}) = \cos(x_1 + x_2)$, (iii) $f(\mathbf{x}) = \cos(x_1 x_2)$.

4.4   Using (4.6) calculate the values of $P^{(t)}$ for $t = 1, 2, 3, ..., 10$. Given $P_{max} = 2$ and $P_{min} = 0$.

4.5   Using (4.7) and (4.8) calculate the values of α_t for $t = 1, 2, 3, ..., 20$. You may assume values for $\alpha_{max} = 1$ and $\alpha_{min} = 0.1$. Plot the graph of α_t against t.

CHAPTER 5

The Firefly Algorithm

5.1 INTRODUCTION

The firefly algorithm was developed by Xin-She Yang working in Cambridge in 2008 and published in Yang (2009). It adapted the behavior of firefly swarms to develop an algorithm for optimizing functions with multiple optima. In particular it used the concept of how the brightness of individual fireflies drew them together and a randomness factor to encourage exploration of the solution space. The publication of the firefly algorithm has led to many papers being published on the analysis and modification of the algorithm and its application to many real world problems and there have been many reports of its successful applications.

5.2 DESCRIPTION OF THE FIREFLY INSPIRED OPTIMIZATION ALGORITHM

The key features of this algorithm are based on a specific set of rules developed by Xin-She Yang, given here:

Rule 1 Fireflies are unisexual, so that one firefly will be attracted to all other fireflies;

Rule 2 Attractiveness is proportional to their brightness, and for any two fireflies, the less bright one will be attracted by (and thus move towards) the brighter one; however, the apparent brightness decreases as their distance apart increases;

Rule 3 If there are no fireflies brighter than a given firefly, it will move randomly.

These three rules may be used to create an optimization algorithm with the additional feature that the brightness is proportional to the value of the objective function. The firefly rules can then be turned into steps in the algorithm by generating the positions of an initial population of fireflies, calculating the value of the objective function for each of these fireflies, and then applying these rules for a number of generations.

The following equation allows the updating of the positions, $\mathbf{x}_i$, of the fireflies as they move around the region taking into account the key factors described above:

$$\mathbf{x}_i^{(t+1)} = \mathbf{x}_i^{(t)} + \beta_0 e^{-\gamma d_{i,j}^2}(\mathbf{x}_i^{(t)} - \mathbf{x}_j^{(t)}) + \alpha \mathbf{r}_i \qquad (5.1)$$

where t denotes the iteration or generation number of the process, $d_{i,j}$ is the distance between any of the pairs of fireflies i and j, $\mathbf{r}_i$ a vector chosen from a uniform or normal distribution and γ a user set parameter usually of $O(1)$. This equation can be interpreted as including two basic elements for the modification of the firefly position values. The term $\alpha \mathbf{r}_i$ allows the random exploration of the region and the extent of its influence can be modified by adjusting the parameter α and the term:

$$\beta_0 e^{-\gamma d_{i,j}^2}(\mathbf{x}_i^{(t)} - \mathbf{x}_j^{(t)})$$

provides control of the convergence of the fireflies towards each other and ultimately the convergence to a specific point. As the distance between the fireflies decreases the exponential term will approach 1. But when the distance between fireflies increases then this term decreases because the power in the exponential function becomes increasingly negative and the exponential term becomes smaller. The overall effect of this term can be controlled by the constant β_0. The value of $\mathbf{r}_i$ is selected from a random distribution which maybe a uniform distribution, adjusted to be in the range $[-1, 1]$ or the standard normal distribution may be used. The value of α is adjusted at each iteration using (5.2).

$$\alpha^{(t+1)} = C\alpha^{(t)} \tag{5.2}$$

Here t is the iteration number and it has been suggested that C may be defined by:

$$C = (10^{-4}/0.9)^{1/n_{gen}} \tag{5.3}$$

where n_{gen} is the maximum number of generations.

This form of C has been found to be effective in many studies. One suggested alternative is to take $C = 0.97$ and other values have been suggested. It is important to keep in mind that larger values of α encourage exploration but may slow down convergence and the choice of C may be problem dependent. For large values of n_{gen} this value of C will tend to 1, for example if $n_{gen} = 100$, $C = 0.9130$, if $n_{gen} = 1000$, $C = 0.9909$. Since $C < 1$ the value of α will be reduced at each iteration decreasing the effect of the random element but this may occur gradually.

The light intensity of the fireflies is proportional to the value of the objective function. Thus the light intensity is computed from the value of the objective function for the current position of the firefly. We wish the firefly to move towards a better objective function value: a lesser value if we are minimizing, which would correspond to the brighter firefly, and the larger value if we are maximizing.

These principles are used in the relatively simple algorithm given by the following pseudo code, where n_{pop} is the number of fireflies:

Algorithm 3 Firefly.

---

for a fixed number of generations **do**
    for $i = 1 : n_{pop}$ **do**
        for $j = 1 : n_{pop}$ **do**
            if brightness of firefly i < brightness of firefly j **then**
                move firefly i towards j; the better value using (5.1)
            end if
            update new solution and update all light intensity values by using the objective function.
        end for
    end for
    Rank fireflies according to fitness; i.e. the objective or fitness function value.
    Store the current best value.
end for

The literature reports that this algorithm has had considerable success for a range of standard global optimization test problems. The algorithm has been extensively used and studied and many suggested improvements have been proposed and in the next section we consider some of these suggested modifications.

5.3 MODIFICATIONS TO THE FIREFLY ALGORITHM

The first proposed modification to the basic firefly algorithm we shall consider was introduced by Fister et al. (2014). This paper considers the randomization process which is a fundamental part of the firefly algorithm. Recall (5.1). The last term of this equation, which promotes the exploration of the region of the global optimization problem, has the form $\alpha \mathbf{r}_i$ where $\mathbf{r}_i$ is the randomization term drawn from an appropriate distribution. For the standard firefly algorithm the distribution used is usually a uniform random distribution with values in the range [0, 1] which is then adjusted to the range [−1, 1] or a standardized normal distribution represented by r_n. Here r_n denotes the normal distribution with mean zero and standard deviation one.

In the paper of Fister et al. different choices for the selection of the randomization parameter are studied to see how they affect the efficiency and accuracy of the firefly algorithm. Specifically they study the effects of using the following distributions:

The uniform distribution This well known distribution is described in Chapter 1 and is used in the standard firefly algorithm

The Normal or Gaussian distribution This distribution is described in Chapter 1 and is also used in versions of the standard firefly algorithm

Lévy flights Lévy distribution is one we have discussed in relation to the cuckoo search algorithm and is described in more detail in Chapter 1

The asymmetric Tent chaotic map This distribution, referred to by Fister et al. (2014) as the Kent map, is defined by:

$$x_{n+1} = \begin{cases} x_n/m, & 0 < x_n \leq m \\ (1-x_n)/(1-m), & m < x_n < 1 \end{cases} \tag{5.4}$$

where $0 < m < 1$. Fister et al. (2014) used $m = 0.7$ in their experiments. The iteration (5.4) has the property that if the initial value of x lies in the interval $[0, 1]$ then all subsequent values of x lie in this interval.

The logistic chaotic map Chaotic maps are implemented from their standard definitions. For the logistic map this has the form:

$$x_{n+1} = rx_n(1 - x_n) \text{ for } n = 0, 1, 2, ... \tag{5.5}$$

The value of r is a specific constant which lies between 0 and 4 and x_0 lies in the interval $[0, 1]$. In the cases where $r = 2$ and $r = 4$ and exact solution is known and is given in (5.6) and (5.7).

$$\text{for } r = 4, x_n = \sin^2(2^n\theta\pi) \text{ where } \theta = \sin^{-1}(x_0^{1/2})/\pi \text{ and } x_0 \in [0, 1] \tag{5.6}$$

$$\text{for } r = 2, x_n = (1/2)\left(1 - (1 - 2x_0)^{2^n}\right) \text{ where } x_0 \in [0, 1] \tag{5.7}$$

A distribution designated by the term RSiTFC This distribution, which comes from the study of star formation in fractal clouds is derived from random sampling using the astronomical concept of the initial mass function. Fister et al. call this Random Sampling in Turbulent Fractal Clouds (for short RSiTFC) and they provide code for the implementation of the method. This distribution has many peaks and valleys and may be useful for strongly multi-model problems because it more successfully models the rapid changes in such varied landscapes.

The chaotic maps are ones where unpredictable results may arise from relatively simple deterministic processes. Both of the chaotic maps can be easily implemented as computer programs and present unpredictability, rather the randomness. Fister et al. (2014) have performed extensive testing on many standard global optimization test problems with dimensions in the range 10 to 50, using each of these methods described and have reported their results and ranked each method using Friedman tests (Friedman, 1937, 1940). Fister et al. (2014) concluded that for each of the dimension ranges the firefly algorithm with Random Sampling in Turbulent Fractal Clouds, called the fractal firefly algorithm, is the most successful in solving a wide range of standard test problems.

The results of Fister et al. (2014) showed significant improvement using alternative probability distributions to the basic firefly algorithm with the Gaussian or normal distribution. They report the results of a study comparing the performance of the fractal firefly algorithm (FFA) with the normal firefly algorithm (NFA), the artificial bee colony algorithm (ABC), the bat algorithm (BA) and differential evolution (DE). They report that the ABC algorithm provides the best overall results but the FFA algorithm is a close second and is significantly better than the NFA and the other algorithms.

The second paper we consider is by Cheung et al. (2014). These authors propose major changes to the original firefly algorithm, in particular adaptive changes to key parameters of the algorithm α, which controls the influence of the random element of the algorithm and the parameter γ which effects the clustering of fireflies through the varying brightness of fireflies and consequently their attractiveness. In addition Cheung et al. propose two new alternative updating rules which are randomly selected and which in turn use a factor that promotes diversity based on Grey coefficients. The aim of this feature is to provide a good balance between the exploration and convergence aspects of the algorithm. We now provide a more detailed description of these improvements to the basic algorithm. We recall the term regarding attractiveness of fireflies from (5.1)

$$e^{-\gamma r_{i,j}^2}(\mathbf{x}_i^{(t)} - \mathbf{x}_j^{(t)}) \tag{5.8}$$

The γ value has a major influence on this term, however the basic firefly algorithm sets this as a constant value $\gamma = 1/\sqrt{L}$ where L is measure of the scale of the problem. The paper of Cheung et al. (2014) proposes that the value of γ is varied as the algorithm proceeds adapting to the current situation. Since this attractiveness is based on the distance between fireflies they propose using the varying distance to calculate the current value of the adapted γ.

A measure of the average distance of the ith firefly to all other fireflies is given by:

$$d_i = \sum_{j=1, j\neq i}^{n_{pop}} \left(\frac{\sqrt{\sum_{k=1}^{n_{var}}(x_{ik} - x_{jk})^2}}{n_{pop} - 1} \right) \tag{5.9}$$

for each $i = 1, 2, ..., n_{pop}$. Here k is the variable component index, n_{var} the dimension of the problem and n_{pop} the size of the population of fireflies. Cheung et al. introduce a distance ratio which can be calculated from:

$$D = \frac{d_{opt} - d_{min}}{d_{max} - d_{min}} \tag{5.10}$$

where d_{max} is the max value of d_i for $i = 1, 2, ...n_{pop}$ and d_{min} the minimum value of d_i for $i = 1, 2, ...n_{pop}$ and d_{opt} is the distance to the optimum current point. Then Cheung et al. introduce an adaptive brightness coefficient which replaces γ, denoted by γ_{ad} defined by

$$\gamma_{ad} = \frac{1}{(1 + \sigma e^{-\rho D})} \tag{5.11}$$

Here D is as defined in (5.10) and Cheung et al. designate ρ and σ as the contraction index and the amplitude factor respectively. In (5.11), if D is large and ρ positive then γ_{ad} tends to 1, however if D tends to zero then γ_{ad} tends to $1/(1+\sigma)$. The γ in the term (5.8) is replaced by γ_{ad} and β replaced by D and we have the new adaptive attractiveness factor:

$$\tau = D\exp(-\gamma_{ad}r^2)$$

Cheung et al. introduce a further parameter calculated using Grey relational analysis designated by η which provides a measure of the diversity of the population. The reader is referred to Cheung et al. for the details of this. Cheung et al. now introduce new updating formulae using the parameters η and τ which bring together the concepts of the adaptive parameter and diversity as follows:

$$\mathbf{x}_i^{(t+1)} = \begin{cases} \mathbf{x}_i^{(t)} + \tau(\mathbf{x}_j^{(t)} - \mathbf{x}_i^{(t)}) + \alpha^{(t+1)}\mathbf{R}_i & \text{if } r_u > 0.5 \\ \frac{(n_{gen}-t)}{n_{gen}}(1-\eta)\mathbf{x}_i^{(t)} + \eta\mathbf{x}_{opt} & \text{otherwise} \end{cases} \tag{5.12}$$

where t is the current iteration and r_u is a random number chosen form a uniform distribution in the range [0, 1]. The alternative formulas being selected randomly the first alternative is very similar to the original firefly equation except adapted parameters are used. Here $\mathbf{R}_i$ is a vector of random values which Cheung et al. (2014) defined by

$$\mathbf{R}_i = \mathbf{r}_i(r_{max} - r_{min})$$

where $\mathbf{r}_i$ is a vector selected from a uniform random distribution in the range [0, 1] with a vector of [0.5, 0.5, ..., 0.5] subtracted from it to provide a random change of sign, n_{gen} is the maximum number of iterations and x_{opt} is the current best value. The constants r_{max} and r_{min} provide the bounds of the solution space. The parameters r_{max} and r_{min} may be replaced by vectors if there are different bounds for the variables of the problem.

The final element in the changes to the algorithm proposed by Cheung et al. (2014) relates to the randomization parameter $\alpha^{(t)}$ we recall from (5.2) that this parameter is updated using the equation:

$$\alpha^{(t+1)} = C\alpha^{(t)}$$

This parameter is a critical one since it governs the emphasis placed on the exploration step and Cheung et al. suggest a range of five alternative updating formulas.

We now describe three of their suggested alternatives, each of the formulas uses a parameter M which is defined by:

$$M = -\exp(-p) \text{ where } p = (n_{gen} - t)/t \tag{5.13}$$

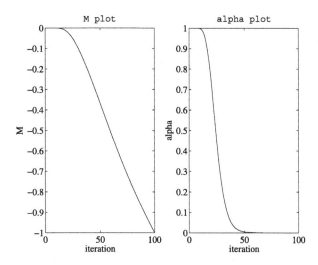

Figure 5.1 Illustrating how α and M vary as iterations increase.

We can now define three of the adaptive formula for $\alpha(t)$ in (5.12), (5.13) and (5.14).

$$\alpha(t) = (e^\delta p\epsilon)^M \qquad (5.14)$$

Here ϵ is set at 10^{-6} and δ is a user defined constant. The values of $\delta = [1, 3, 5, 7, 9]$ have been tested by Cheung et al. (2014) for (5.14) and the values $\delta = [10, 30, 50, 70, 90]$ for (5.15) and (5.16).

$$\alpha(t) = \exp\left(\delta p^{1/t}\right)^M \qquad (5.15)$$

$$\alpha(t) = (e^\delta t^{-n_{pop}/n_{var}})^{\frac{M(n_{gen}-t)}{n_{gen}}} \qquad (5.16)$$

Here n_{pop} is the population size and n_{var} is the number of variables in the problem.

All of these formulas will provide a monotonic reduction in the value of $\alpha(t)$ but in significantly different ways. They provide different methods which may be more suitable for specific problems.

Making use of (1.14), Figure 5.1 illustrates how α and M vary as the iterations increase.

The authors of this paper, Cheung et al., provide extensive numerical studies of the performance of their adaptive algorithm on a number of standard global optimization test problems and conclude that their algorithm and its variants performance ranks above the performance of the standard firefly algorithm, the standard particle swarm optimization, adaptive particle swarm optimization and Grey particle swarm optimization in both efficiency and accuracy. The ranking of the performance of each modified algorithm is established using Friedman test.

Table 5.1 Showing variation in performance of the firefly algorithm with swarm size

Swarm Size	Mean	Best	Worst	St Dev
10	1.8982	2.2289×10^{-4}	9.4892	3.8938
20	0.4748	7.5936×10^{-5}	9.4887	2.1217
30	3.1320×10^{-4}	6.7564×10^{-5}	5.9843×10^{-4}	1.3804×10^{-4}
40	2.2990×10^{-4}	5.0144×10^{-5}	7.0233×10^{-4}	1.6851×10^{-4}

Table 5.2 This shows how the performance of the algorithm varies with increasing number of iterations

Iterations	Mean	Best	Worst	St Dev
200	2.3353	2.7351×10^{-4}	9.4892	4.1515
500	0.4749	1.5534×10^{-4}	9.4847	2.1216
1000	3.2622×10^{-4}	9.6243×10^{-5}	8.0176×10^{-4}	1.9406×10^{-4}
2000	1.7715×10^{-4}	7.5049×10^{-6}	3.6342×10^{-4}	1.0191×10^{-4}

5.4 SELECTED NUMERICAL STUDIES OF THE FIREFLY ALGORITHM

There are many numerical studies that can be performed regarding the key parameters of the firefly algorithm we have selected a number of these and the results are given below.

The studies are carried out using the parameters population size $n_{pop} = 30$, $\beta_0 = 1$, $\gamma = 1/\sqrt{L}$ and updating factor for α as $(10^{-2}/0.9)^{1/n_{gen}}$, the initial value of α is taken as $0.1L$. These are only changed in experiments relating to them, the value of L is taken as the maximum range of the variables of the problem minus the minimum value of the range. Twenty runs of the algorithm are performed for each study for optimizing the egg-crate problem with four variables. The number of iterations is fixed at 1000. The first experiment examines the effect of population size we take swarm size = 10, 20, 30, and 40 and the results are shown in Table 5.1.

The results of Table 5.1 show significant improvement in the performance of the algorithm with increasing swarm size.

We now examine how the performance of the algorithm is effected by the number of iterations performed. In general Table 5.2 shows that the performance of the algorithm improves with increasing number of iterations, although the improvement from 1000 to 2000 iterations is small.

We now consider the influence on the behavior of the algorithm of changes in the key parameters β_0, γ and the manner in which α is adjusted. We consider first the parameter γ, using 1000 iterations, the results are given in Table 5.3. These tests show that for the egg-crate problem with four variables, this limited range of values of γ has little influence on the results.

We now consider the effect of changes in the value of the parameter β_0, which modifies the attractiveness factor, in Table 5.4. In Table 5.4 we see significant differences

Table 5.3 Showing change in performance of the algorithm with changing values of γ

γ	Mean	Best	Worst	St Dev
0.1	2.8630×10^{-4}	2.7064×10^{-5}	8.6153×10^{-4}	2.2890×10^{-4}
1	3.7017×10^{-4}	5.8220×10^{-5}	6.9999×10^{-4}	1.7726×10^{-4}
$1/\sqrt{L}$	3.7885×10^{-4}	1.6845×10^{-4}	6.9308×10^{-4}	1.2832×10^{-4}

Table 5.4 This shows the performance of the firefly algorithm with changes in the value of β_0

β_0	Mean	Best	Worst	St Dev
2	0.0112	0.0023	0.0304	0.0071
1	0.0356	5.0726×10^{-4}	0.7053	0.1576
0.5	2.7926×10^{-4}	3.5878×10^{-5}	5.4335×10^{-4}	1.3947×10^{-4}
0.2	5.9521×10^{-4}	4.6238×10^{-5}	0.0013	3.1703×10^{-4}

Table 5.5 Showing the results of using alternative updating formula for α. The last two values vary dependent on the maximum number of generations used

C	Mean	Best	Worst	St Dev
0.97	0.9490	3.2535×10^{-27}	9.4882	2.9203
0.2	19.3926	1.3765	39.4851	9.4964
$(10^{-4}/0.9)^{(1/n_{gen})}$	0.9488	7.2968×10^{-9}	9.4882	2.9204
$(10^{-2}/0.9)^{(1/n_{gen})}$	2.7409×10^{-4}	1.5366×10^{-4}	5.5453×10^{-4}	1.0216×10^{-4}

in the performance of the algorithm with different values of β_0. In particular the smaller values of β_0 provide significantly better results.

In Table 5.5 we consider how the performance of the algorithm is changed by using different forms of the updating formula for α, we recall, this takes the form: $\alpha^{(t+1)} = C\alpha^{(t)}$. We take $C = 0.97$, $C = 0.2$, $C = (10^{-4}/0.9)^{1/n_{gen}}$, n_{gen} is the maximum number of generations. These results are shown in Table 5.5.

Table 5.5 shows the results are very poor for the value 0.2 but much better results are obtained for the values 0.97 and $C = (10^{-4}/0.9)^{1/n_{gen}}$. The result obtained using a the value of $C = (10^{-2}/0.9)^{1/n_{gen}}$ appears to be the best of all. However occasionally the algorithm is trapped before the optimum is reached due to random fluctuations leading to an occasional unusually poor result for $C = (10^{-4}/0.9)^{1/n_{gen}}$. In fact since the values for 10,000 and 1000 generations are 0.9991 and 0.9909 respectively, there is very little difference between them. Since the value of this parameter may be critical, to confirm any real improvement further studies should be performed.

As a final test of the firefly algorithm we consider the effect of changing the random number distribution from the normal distribution to The Lévy distribution. There has been a great deal of discussion about the improvements that may be achieved using the

Table 5.6 Minimization of the Egg crate function with 10 variables compare the use of Levy and Normal distributions

	Mean	Best	Worst	Standard Deviation
Lévy	3.7101×10^{-4}	2.88164×10^{-4}	4.6744×10^{-4}	6.7993×10^{-4}
Normal	8.5395	4.4437×10^{-4}	18.9764	5.3858

Table 5.7 Minimization of the Egg-crate function with 10 variables compare the results using Lévy parameter values 1 and 1.5

c	Mean	Best	Worst	Standard Deviation
1	3.7101×10^{-4}	2.88164×10^{-4}	4.6744×10^{-4}	6.7993×10^{-4}
1.5	0.9441	1.5358×10^{-4}	9.4482	2.0003

Lévy distribution, see Fister et al. (2014). It has been suggested that the Lévy distribution would allow a more thorough exploration of the search space for strongly multi-model functions. We have used the Lévy distribution in Chapter 4 and discussed it in some detail in Chapter 1. To test this proposition we optimize the egg crate problem with 10 variables with the normal and the Lévy distributions. The results are given in Table 5.6 for 10000 iterations. In each case 10 runs of the algorithm are used.

Table 5.6 shows that much better results are achieved using the Lévy distribution. However the problems chosen are selected to test the explorative properties of the Lévy distribution and these good results may not be reflected in other problems with different dimensions and different complexities. In addition the efficiency of the method may not have improved in terms of computational effort and time taken to obtain the required solution. These results certainly merit further study on a wider range of test problems.

We have seen in Table 5.6 how using the Lévy distribution can change the performance of the Firefly algorithm for certain optimization problems. We now discuss how changing a parameter in the definition of the Lévy distribution may produce further changes in the performance of the algorithm.

The general form of the Lévy distribution is controlled by a specific parameter which we have denoted by c which alters the shape of the Lévy distribution, see Chapter 1 for a more detailed discussion. We now provide Table 5.7 which shows how performance varies with changes in the value of c, where $c > 0$. The same values for the parameters that we have used in Table 5.6 have been used in this experiment. Table 5.7 shows that significant improvement is achieved setting $c = 1$ rather than $c = 1.5$. However further studies would be needed to confirm this result, but it does indicate this factor may effect the performance of the algorithm and consequently needs further study using a range of values for the parameter c.

We now provide graphs giving an overall picture of the convergence of the algorithm and how the firefly swarm behaves in converging to the global optimum. In Figure 5.2 we show the performance of the algorithm for Rosenbrock's function which has only

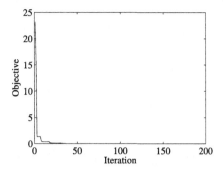

Figure 5.2 Convergence for Rosenbrock's function with two variables.

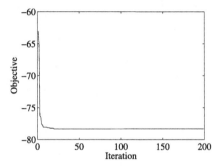

Figure 5.3 Convergence for Styblinski-Tang function for two variables.

one minimum point. There is only limited information that can be gleaned from a single run of the algorithm but the figure does show a rapid convergence to the correct minimum value.

Figure 5.3 shows rapid convergence to the correct minimum of the Styblinski-Tang function. There is a hint of a slower convergence where the process searches round a specific value without changing.

Figure 5.4 shows the convergence to the correct value for the minimum of Rastrigin's function. This function is a more challenging one with many local minima and there is an indication of slower convergence in the process where the process searches round a specific value without changing the minimum. To examine this process further we consider the performance of the algorithm on a series of problems; the egg-crate problem with four and six variables.

Figure 5.5 shows a rather slower approach to the optimum for the four variable case and exhibits some iterations searching around the same point. This is to be expected since the size of the search region increases rapidly with the number of variables in the problem.

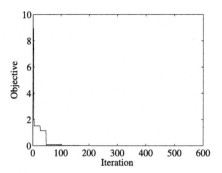

Figure 5.4 Convergence for Rastrigin's function with two variables.

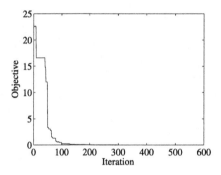

Figure 5.5 Convergence for egg-crate function with four variables.

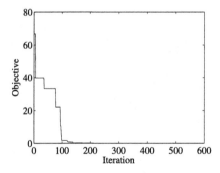

Figure 5.6 Convergence for egg-crate function with six variables.

Figure 5.6 shows an even slower approach to the optimum for the six variable case and exhibits some iterations searching around the same point. This is expected since the size of the search region increases rapidly with the number of variables in the problem.

The next set of graphs show the behavior of the firefly swarm on contour graphs of the Rosenbrock function, the Styblinski-Tang function and the egg-crate function.

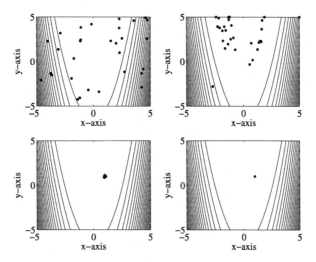

Figure 5.7 Showing convergence of swarm to minimum of Rosenbrock's function, 0 at [1, 1], at 0, 10, 200 and 500 iterations. Swarm size 30.

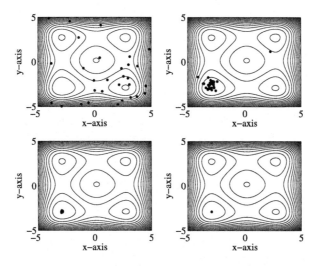

Figure 5.8 Plot of the Styblinski-Tang function at [−2.9035 − 2.9035] for 0, 50, 200 and 500 iterations. Swarm size 30.

Figure 5.7 shows the rapid convergence of the firefly algorithm for optimizing the Rosenbrock function to the optimum point by the 500th iteration. The swarm has gathered close together at the same point.

Figure 5.8. For the Styblinski-Tang function there are four minimum points but this shows the rapid convergence of the firefly algorithm to the global minimum point by the 500th iteration the swarm has gathered close together at the same point.

5.5 DEVELOPMENTS OF THE FIREFLY ALGORITHM

In a recent paper Yang and He (2013), provide some theoretical insight into the performance of the Firefly algorithm with the hope of improving the balance between theoretical justification and actual performance. In particular the paper provides some numerical evidence of how the firefly algorithm produces an effective balance between exploration and exploitation. We discuss this paper in this section. Yang et al. consider a landscape based strategy for analyzing the progress of an algorithm, clearly different problems will have different landscapes and therefore performance will be problem dependent. However the central question for search algorithms is the balance of time spent between exploration and exploitation, where exploitation refers to the process of converging to an acceptably accurate solution. Yang et al. modeled the process of the search for a global optimum in a multi-modal landscape as a diffusion process. Using random walks within a small region radius a in a much larger region radius b, they showed that a formula may be derived for the two dimensional case which provides the optimal balance between exploration and exploitation. According to Yang and He, this takes the form

$$t_{opt} = t_a/(t_b^2) = D/(a^2(2 - 1/\log_e(b/a))^2)$$

Here t_a and t_b are the time spent on exploitation and exploration parts searching through the region respectively.

In this formula D is called the diffusion coefficient and governs how the diffusion takes place throughout the solution space and can be compared to a random walk so Yang et al. state that D may be replaced by $s^2/2$ where s is the step length, taking the ratio of b to a as 10. Yang et al. state that this leads to an optimum ratio of approximately 0.2 for the balance between exploration and exploitation, this implies considerably more time should be spent on the exploration process. Bearing this result in mind, Yang et al. perform a series of numerical studies using the firefly algorithm and count the number of function evaluations used in the exploration and exploitation phases. The results they obtain verify that the best ratio is 0.2 which corresponds to the theoretical result. Thus, they state that this gives the firefly algorithm's actual performance some consistency with the theoretical result. Yang et al. also discuss how this result may be extended to higher dimensions.

5.6 SOME APPLICATIONS OF THE FIREFLY ALGORITHM

Cheung et al. (2014) report they have used the Firefly algorithm to provide highly accurately and robust predictions of the helix structure of the tertiary structure of proteins.

In a paper by Abdelaziz et al. (2015) the authors provide a useful review of the FFA and its applications. Amongst these they describe the Nuclear Reactor Reload Problem which relates to the efficient scheduling of the replacement of part of the fuel of a nuclear reactor. This produces a complex multi-objective function problem since competing objectives must be satisfied. In addition they cite the CSA application to electrical load forecasting, a complex task since it is subject to the vagaries of consumer demand.

Reddy and Chandra-Sekhar (2014) apply the Firefly Algorithm to the problem of optimizing the combined economic load scheduling and total fuel costs, but at the same time minimizing emission production to reduce the environmental impact. This leads to a multi-objective function problem.

5.7 SUMMARY

In this chapter we have described the basic firefly algorithm and some amendments suggested by workers in the field; there have been many of publications relating to the firefly algorithm and the selection of representative papers has been difficult. However our selection does deal with important areas such as the selection of the random distributions to obtain the best exploration of the solution space, the best choice for the for adaptive formula for updating key parameters. We also provide a range of numerical studies of key parameters affecting the performance of the algorithm such as swarm size and updating techniques for the α parameter. This is not a research paper and the studies are not exhaustive but it is hoped they give an indication of the process used to obtain an insight into the behavior of the methods.

In addition we provide a discussion due to Yang and He (2013) of how the firefly algorithm deals with the key issue of obtaining the best balance of effort spent on the explorative and exploitative elements of algorithm.

5.8 READER EXERCISES

5.1   Consider the updating formula for the parameter α given by (5.13) and (5.14). For the updating formula (5.14):

$$\alpha(t) = ((e^{\delta}(n_{gen} - t))/t\epsilon)^M$$

draw graphs of α against t for $t = 1$ to 200 in steps of 1 and $n_{gen} = 200$ for values of the parameter $\delta = 1$, 3, 5 and $\epsilon = 1$.

5.2   For the formula (5.16) for updating α:

$$\alpha(t) = (e^{\delta} t^{-n_{pop}/n_{var}})^{\frac{M(n_{gen}-t)}{n_{gen}}}$$

take $\delta = 3$ and $n_{pop} = 20$, and plot graphs of the behavior of this formula for values of $n_{var} = 2$, 5, 10. The graphs should be plotted for $t = 1$ to 200 in steps of 1 and $n_{gen} = 200$.

5.3 Use formula (5.9) to find the average distance between a population of firefly located at $[1, 1]^T$, $[2, 2]^T$, $[1, 2]^T$, $[3, 2]^T$ and $[3, 3]^T$, labeled 1, 2, 3, 4 and 5 respectively. Calculate the value of D. Use the objective function $x_1^2 + x_2^2$ to determine the required d_{opt} value.

5.4 Using the logistic map, see (5.5) and taking $x_0 = 0.25$ for (5.6) and $x_0 = 0.1$ for (5.7), generate the first 5 values in the sequence for x in the cases where $r = 2$ and $r = 4$ using the standard logistic equation. Check your results by using the formulae given by (5.6) and (5.7).

5.5 Write a MATLAB script or a program in any other language to generate points using (5.6) and plot them over 200 generations. Take $x_0 = 0.1$.

5.6 Use the asymmetric Tent map to generate a sequence of values for x using the value $m = 0.7$. Generate 10 values and plot the values you have generated. Take $x_0 = 0.1$.

CHAPTER 6

Bacterial Foraging Inspired Algorithm

6.1 INTRODUCTION

A key element of all animal life is the continuous process of foraging for food. This is an obsession for many forms of life, only interrupted when satiated and by the need for rest. However the process of foraging must be carried out in sufficiently efficient manner so that the energy expended in gaining the required nutrients is less than energy added by the consumption of the food; thus the search for food should be carried out in an optimal way. Passino (2002) studied foraging in relation to bacterial behavior and as a result produced the Bacterial Foraging Optimization Algorithm (BFOA). In Section 6.2 we give a formal description of this method.

6.2 DESCRIPTION OF THE BACTERIAL FORAGING OPTIMIZATION ALGORITHM

The algorithm is not a precise implementation of the bacterial foraging process but simulates some of the key features of the behavior of bacteria in the development of the optimization algorithm. These features are:

Chemotaxis. This simulates the process of movement of a bacterial cell. Ecoli, for example, move by a mixture of tumbling and swimming, which these cells can achieve by means of the various movements of spidery appendages called flagella. This allows the bacterium to move through the nutrient material with the aim of gaining the optimum benefit. There are two forms of movement, the purely random movement of tumbling and the directed process of swimming which takes random steps but ones following the changes in chemical concentrations. This allows the bacterium to achieve its key aim which is to reach a higher density of nutrition. This provides an improvement in the fitness of the bacterium. These two classes of step reflect the exploration and exploitation features so useful in global optimization algorithms.

Swarming. This simulates a feature that groups of bacteria exhibit: their tendency to swarm in the form of a concentric ring, each releasing an attractant for the others when stimulated by the presence of the nutrient. Thus, they move together to sweep up the nutrient for their mutual benefit by following the upward nutrient gradient. However, bacteria also emit chemical signals that deter other bacteria, consequently there is a balance of attractive and repellent effects which are reflected in the link between bacteria.

Introduction to Nature-Inspired Optimization
DOI: 10.1016/B978-0-12-803636-5.00006-2

Reproduction. The least healthy bacteria die but the healthy bacteria reproduce by splitting into two. The swarm size of the bacteria is thus kept constant.

Elimination and dispersal. Occasionally significant changes may affect the whole bacterial population due to, for example, a rise in temperature when a group of bacteria may die or be dispersed to another region. This process can be simulated by removal at random of some bacteria and their random replacement throughout the region in which the problem is defined, boosting the exploratory aspect of the process.

The algorithm implementation has more complexity than some other biologically inspired algorithms which presents challenges in studying the performance of the method and its efficient implementation. Having described the basic elements of the algorithm, these elements must be simulated in a programmable way. This is achieved by setting up a series of nested loops for the four processes previously described, implementing the movement processes tumbling and swimming and introducing an objective or fitness function to test the fitness of the bacteria.

The process of tumbling has been implemented by Passino using the following equation which adjusts the position vector $\mathbf{x}_i$ of the bacterium i using (6.1)

$$\mathbf{x}_i^{(j+1,k,l)} = \mathbf{x}_i^{(j,k,l)} + C_i \frac{\mathbf{u}_i}{\sqrt{\mathbf{u}_i^{\mathsf{T}} \mathbf{u}_i}} \tag{6.1}$$

Here the index j refers to the index of the chemotaxis step, k the index of the reproduction loop and l the elimination and dispersal loop. The symbol $\mathbf{u}_i$ represents a vector of random values uniformly selected in the range $[-1, 1]$ and C_i the step size for the specific bacterium i. The other mode of movement for the bacteria is the swimming process and this can be simulated by the swimming pseudo code, given in **Algorithm 4**.

Algorithm 4 Bacteria Swimming.

---

if $J(i, j + 1, k, l) < J_{last}$ **then**
    keep the best so far:
    $J_{last} = J(i, j + 1, k, l)$
    $\mathbf{x}_i^{(j+1,k,l)} = \mathbf{x}_i^{(j,k,l)} + C_i \mathbf{u}_i / \sqrt{\mathbf{u}_i^{\mathsf{T}} \mathbf{u}_i}$
    Use this new $\mathbf{x}_i$ to compute a new value of the objective function J.
end if

Although in this algorithm the same step adjustment as given by (6.1), its use only occurs if there is an improvement in the objective function value. This is given by $J(i, j + 1, k, l)$, so swimming steps are directed towards improvement in the value of the objective function. These swimming steps are performed a specific number of times, denoted by n_s preset by the user.

The tendency for the bacteria to swarm together in the direction of increasing levels of nutrients is simulated by the expression J_{cc} which shows the interaction of any bacteria $x^*(j, k, l)$ with the other bacteria of the population $i = 1, 2, .., n_{pop}$ where n_{pop} is the number of bacteria:

$$J_{cc}(x^*) = \sum_{i=1}^{n_{pop}} \left[-d_a \times \exp \left(-w_a \sum_{m=1}^{n_{var}} (x_m^* - x_m^i)^2 \right) \right]$$
$$+ \sum_{i=1}^{n_{pop}} \left[h_r \times \exp \left(-w_r \sum_{m=1}^{n_{var}} (x_m^* - x_m^i)^2 \right) \right] \quad (6.2)$$

Note that some of the indices are dropped for clarity. Here n_{var} is the dimension or number of variables of the problem, and m is the dimension index. d_a and w_a are attractant factors and h_r and w_r repellent factors. These values are decided by the user to enhance the performance of the algorithm.

It is interesting to interpret the behavior of this function for example if all points are close together then the terms in the exponential tend to zero and hence the exponential of these values tends to 1 so the value of J_{cc} tends to $-d_a \times n_{pop} + h_r \times n_{pop}$ or $n_{pop}(h_r - d_a)$. However when the points are distant from one another the terms in the exponent are large and negative consequently the exponent term tends to zero so the influence of the J_{cc} term dies away. How rapidly this occurs depends on the values of w_a and w_r. We also note that the second term provides a repulsion factor which tends to disperse the bacteria. The relative strength of these terms is controlled by the constants h_r and d_a. This J_{cc} quantity is added to the current objective function J to introduce the effect of swarming.

Using (6.1), and (6.2) and the swimming pseudo code we can write the pseudo code for the complete algorithm. The implementation of the algorithm, given by **Algorithm 5**, takes place within a succession of nested loops. This pseudo code provides the basis for an implementable algorithm.

The algorithm has several parameters which must be set carefully. This of course presents significant difficulties in choosing the most efficient combination of parameters in the algorithm for a specific problem. In particular since we are dealing with nested loops the number of inner iterations and computations can rise rapidly. For example if we take $n_c = 70$, $n_{re} = 5$, $n_{ed} = 20$ this implies inner iterations of 7000. Taking account that each of these iterations may involve actions for each bacterium then the process will involve a large number of computations. Consequently great care must be taken in the selection of these parameters. In addition to loop counts the other constants have to be set sensibly and may be problem dependent.

There have been many modifications to the basic algorithm with the aim of improving its performance. These include the adaptive choice of some of the parameters

---

Algorithm 5 BFO.

STEP 0: Initiate all constants

$n_{pop} \leftarrow$ number of bacteria

$n_s \leftarrow$ number of swim steps

$n_{var} \leftarrow$ dimension of the problem

$n_c \leftarrow$ number chemotaxis steps

$n_{re} \leftarrow$ number of reproduction steps

$n_{ed} \leftarrow$ number of elimination and dispersal steps

$p_{ed} \leftarrow$ probability of elimination and dispersal

$C \leftarrow$ step vector

Generate randomly an initial population of bacteria within the region of the problem definition. Then execute the nested loops.

for $l = 1 : n_{ed}$ **do**

    for $k = 1 : n_{re}$ **do**

        for $j = 1 : n_c$ **do**

            for $i = 1 : n_{pop}$ **do**

                Compute for each bacterium the objective function $J(i, j, k, l)$

                Calculate $J(i, j, k, l) = J(i, j, k, l) + J_{cc}$.

                Where J_{cc} is calculated using (6.2)

                Thus the new J value reflects the effect of swarming.

                Save the current value of J in J_{last}.

                Perform the tumbling step using (6.1).

                Compute the objective function value for the new point and add J_{cc}.

                Now simulate the bacterium swimming process by

                executing **Algorithm 4** for n_s steps.

                for $s = 1 : n_s$ **do**

                    Execute swimming pseudo code, **Algorithm 4**.

                end for

            end for

        end for

    Reproduction step for the current k and l. Calculate for each member of the bacterial population the health cost value J_{health}^i using

    for $i = 1 : n_{pop}$ **do**

        $J_{health}^i = \Sigma_{j=1}^{n_c+1} J(i, j, k, l)$

    end for

    Here the values of J are summed over all chemotaxic steps, this provides a value for the health costs of each bacterium.

    Sort bacterium in ascending order of their health cost values. It is important to note that higher cost means lower health.

    Select a number of bacteria with highest health cost values to die and the remaining are split and replaced in the parent location.

    end for

With probability p_{ed} eliminate and disperse each bacterium keeping number of bacterium constant by generating new ones randomly.

end for

---

and different schemes to ensure the region in which the problem is defined are properly explored to find the global minimum and that a good level of accuracy is achieved. In the next section we will describe a selection of these amendments and proposed improvements.

6.3 MODIFICATIONS OF THE BFO SEARCH ALGORITHM

There have been many suggested modifications to BFOA we will consider only a selection in this section. An amendment rapidly introduced by Liu and Passino (2002) to their new algorithm was an amendment to the factor J_{cc} which reflected the affects of communication between bacteria. They introduced the factor J_{ar} which takes account of environmental considerations. Here J_{ar} is defined by (6.3).

$$J_{ar} = \exp(M - J(\mathbf{x}))J_{cc} \tag{6.3}$$

where M is an adjustable parameter. Clearly when $J(\mathbf{x})$ tends to M, J_{ar} tends to J_{cc}, however if $J(\mathbf{x})$ is distant from M then, J_{ar} will be different from J_{cc}. The modified objective function then takes the form:

$$J = J + J_{ar}$$

Clearly M must be selected appropriately. If M is large, then J_{ar} will amplify J_{cc}. Then the search space may be dominated by the attractive secretant, but if M is small, J_{ar} is smaller than J_{cc} and the search space is dominated by the effect of the nutrient.

A second amendment to the basic algorithm was introduced by Dasgupta et al. (2009). In this paper Dasgupta et al. considered theoretical aspects of the BFO algorithm and concluded that the BFO algorithm may be considered a modified classical gradient method and they obtained a characteristic equation for the chemotaxis process dependent on the gradient. In addition they suggested that there was a need to modify the basic algorithm since it was possible that the algorithm could come close to the minimum but then keep taking chemotaxis steps and consequently oscillating around it. To avoid this they indicate the step size should be changed adaptively as the algorithm proceeds. In particular the value of the step size C should be modified by using (6.4).

$$C = |J(\mathbf{x})|/(|J(\mathbf{x})| + \lambda) \tag{6.4}$$

Here $J(\mathbf{x})$ is the value of the objective at the current bacterium location and λ is a positive constant. Assuming the global minimum value of the objective function is zero, C will approach zero as the global minimum is approached. The value λ must be selected carefully by the user. An interesting interpretation of this, given by the Dasgupta et al. is that when the objective value is large then bacterium will be in a noxious region and a large step size will allow quicker exploration and the discovery of more nutritional areas. This will be the case since as the objective function becomes larger the value of C will approach 1. Dasgupta et al. suggest an alternative modification which takes the form:

$$C = |J(\mathbf{x}) - J_{best}|/(|J(\mathbf{x}) - J_{best}| + \lambda) \qquad (6.5)$$

where J_{best} is the current objective function value of the global best bacterium. This modification takes account of the bacteria position relative to the location of the current global best bacterium. Dasgupta et al. have carried out extensive studies on a range of standard test problems for global optimization and report good and improved results in comparison to a range of competing techniques. In particular they have found that the adaptive approach using (6.4), the simpler form for C, appears to be the most successful.

The next paper we consider is that of Biswas et al. (2007). In this paper Biswas et al. propose more extensive amendments to the basic BFO algorithm. In particular they integrate the BFO algorithm with some aspects of the methods of Differential Evolution. The resulting algorithmic process they call Chemotactic Differential Evolution (CDE). As ever in this field the aim is to increase efficiency and accuracy i.e. to find the global minima to an acceptable accuracy with the least computation. In Chapter 2 we have described the major features of Differential Evolution and Genetic Algorithms in particular we have described the various forms of mutation, crossover and selection used in both genetic algorithms and differential evolution. Biswas et al. used the Differential Evolution strategy denoted by $DE/rand/1$ (see Chapter 2) to introduce differential evolution mutation to the BFOA this takes the form, using a simplified notation:

$$\mathbf{V}^{(t)}(i, j + 1) = \mathbf{x}(r_1) + F(\mathbf{x}(r_2) - \mathbf{x}(r_3)) \qquad (6.6)$$

Here t is the current iteration, r_1, r_2 and r_3 are randomly selected integer values in the range $[1, n_{pop}]$ where n_{pop} is the bacterium population size. The random numbers r_1, r_2 and r_3 must be selected so that $i \neq r_1 \neq r_2 \neq r_3$. Thus $\mathbf{x}(r_1)$, $\mathbf{x}(r_2)$ and $\mathbf{x}(r_3)$ are the selected points. Note that in (6.6) the loop subscripts have been removed for clarity.

This new vector $\mathbf{V}$ and the current vector $\mathbf{x}$ are used to introduce crossover by exchanging components p with a new vector $\mathbf{U}$, the components of which are calculated by (6.7)

$$U_p^{(t)} = \begin{cases} V_p^{(t)}(i, j + 1) & \text{if } r_u \leq C_{rate} \text{ or } p = r(i) \\ x_p^{(t)}(i, j + 1) & \text{if } r_u > C_{rate} \text{ or } p \neq r(i) \end{cases} \qquad (6.7)$$

where r_u is a uniform random number in the range $[0\ 1]$ and $r(i)$ is a randomly chosen index to ensure $\mathbf{U}$ gets at least one component of $\mathbf{V}$ and C_{rate} determines the crossover probability. The value or C_{rate} is decided by the user.

Selection is implemented as usual by using the objective function to test the newly generated trial vector $\mathbf{U}$ to obtain the best new bacterium vector from components of $\mathbf{x}$

and $\mathbf{U}$. Thus the effects of mutation and crossover are passed on in the selection process to $\mathbf{x}$ via the candidate vector $\mathbf{U}$. So for selection we have:

$$\text{If } J(U^{(t)}(i, j+1)) < J(\mathbf{x}^{(t)}(i, j+1)) \quad \text{then } \mathbf{x}^{(t)}(i, j+1) = \mathbf{U}^{(t)}(i, j+1)$$
$$\text{otherwise no change to } \mathbf{x} \tag{6.8}$$

A few additional changes were suggested by Biswas et al. (2007); an inertial weight was introduced and the step size C adjusted adaptively changing its value dependent on the current value of the objective function using (6.9).

$$C(i) = \frac{[J^{(t)}(i, j)]^{1/3} - 20}{[J^{(t)}(i, j)]^{1/3} + 300} \tag{6.9}$$

The modification for the tumbling and swimming steps is given by (6.10). Here the inertial weight w is introduced:

$$\mathbf{x}_i^{(j+1,k,l)} = w\mathbf{x}_i^{(j,k,l)} + C_i \frac{\mathbf{u}_i}{\sqrt{\mathbf{u}_i^\top \mathbf{u}_i}} \tag{6.10}$$

It should be noted that all the changes are made within the chemotaxis step of the BFO algorithm. Biswas et al. carried out extensive numerical studies using standard global optimization test problems and reported considerable improvements over the performance of the basic BFO algorithm. They also reported some improvements on a differential evolution approach and a genetic algorithm method but here the improvement was less significant. We now carry out some numerical experiments to study the behavior of the basic BFO Algorithm.

6.4 SELECTED NUMERICAL STUDIES OF THE BFO SEARCH ALGORITHM

The basic bacterial foraging algorithm is more complex in structure than many nature inspired optimization algorithms particularly in regard to selecting the values for the various iterations and sub iterations. Here we consider only a selection of the possible studies using the basic implementation of the algorithm. For these studies the values for the various iteration loops are set as $n_c = 30$, $n_s = 5$, $n_{re} = 10$, $n_{ed} = 20$. The other parameters are set as $C = 0.1$, $da = 0.1$, $wa = 0.2$, $hr = 0.1$, $wr = 10$, $p_{ed} = 0.25$ unless otherwise stated for the particular study. Of course these parameters may not provide the best results. Because of the large number of parameters and parameter interactions, care should be taken in the interpretation of the results.

In the first study we consider the number of bacteria which is equivalent to the swarm size. We consider the number of bacteria as 10, 20 and 30 for solving Rastrigin's problem with four variables. Using 20 runs of the algorithm for each number of bacteria we obtain Table 6.1.

Table 6.1 Showing the effect of increasing the number of bacteria used by the BFO algorithm

n_{pop}	Average	Best	Worst	St Dev
10	2.1366	0.9153	3.0421	0.5904
20	1.1475	0.1551	2.0568	0.5249
30	0.8623	0.1340	2.2051	0.4981

Table 6.2 Showing the effect of changes in the parameter p_{ed} on the performance of the BFO algorithm

p_{ed}	Average	Best	Worst	St Dev
0.1	1.0949	0.4068	1.6346	0.3638
0.25	1.2355	0.3576	2.2497	0.5140
1	2.2488	1.1892	3.7222	0.6894

Table 6.3 Showing the effects of the parameter C on the performance of the BFOA

C value	Average	Best	Worst	St Dev
0.1	1.1474	0.1163	2.2126	0.5643
0.25	1.5646	0.2442	2.5290	0.5881
0.4	2.1332	0.9818	3.2256	0.5993

There is gradual improvement in the performance of the method as the number of bacteria are increased. However, the bacteria numbers for this combination of parameters do not appear to be critical. Some researchers have reported using much larger numbers of bacteria for high dimension problems.

The following study examines the effect of using different values of the constant p_{ed}. The value p_{ed} determines the proportion of the original bacteria which are replaced. The study uses a bacteria population of 20 and is tested with the same set of parameters as the previous study. Twenty runs of the algorithm are performed for the values of $p_{ed} = 0.1$, 0.25 and 1 successively and this leads to Table 6.2.

For this parameter set the smaller values of p_{ed} produce a better performance for the algorithm. Taking the value of p_{ed} to be 1 produces a relatively poor performance.

In Table 6.3 we examine the effect changes in the parameter C on the performance of the BFOA. The values we take are $C = 0.1$, 0.25 and 0.4. This controls the step size value within the algorithm. The parameter p_{ed} is reset to 0.25, and remaining parameters are set to the same values as in the other runs. As usual, 20 runs are performed.

This shows better results for the smaller values of C and merits further study with even smaller values of C.

There are a large number of parameters used in the BFOA and consequently we cannot study the effect of each one so we will combine a pair of parameters in the next

Table 6.4 Showing effects in changes in the parameters d_a and w_a on the performance of the BFOA

Set	Average	Best	Worst	St Dev
S1	1.0330	0.1200	1.9566	0.6084
S2	1.4992	0.4288	2.3010	0.5025
S3	1.2664	0.1839	2.2585	0.6765

Table 6.5 Showing the effects on the performance on the of the BFOA of changes in the parameter h_r

h_r	Average	Best	Worst	St Dev
10	1.1860	0.0320	2.1741	0.6765
5	1.4169	0.5657	2.4063	0.5093
1	1.4756	0.4592	2.0278	0.4381

study these are the values d_a and w_a which are associated with the modified objective or fitness function used in the algorithm. We use the pairings $d_a = 0.1$, $w_a = 0.2$; $d_a = 0.2$, $w_a = 0.4$ and $d_a = 0.4$, $w_a = 0.8$ called S1, S2 and S3 respectively. The results are given in Table 6.4.

The results show little difference in the performance of the parameter sets but the smallest values of S1 have produced the best result.

We now consider the effect of changes to the parameter w_r on the performance of the algorithm. In Table 6.5 we consider runs using the values of the parameter $h_r = 10$, 5 and 1. The parameter w_r is held at 0.1. These parameters are also part of the definition of the modified objective function.

The changes in the performance of the algorithm for the different parameters are limited even though the changes in the parameter are relatively large. The value $h_r = 10$ does however produce the best overall result but this may be a random effect and merits further study.

In the final numerical study we shall examine the effect of the number of chemotaxis loop iterations that are performed. As the description in Section 6.2 of this chapter indicates the chemotaxis loop plays an important role in the way the algorithm functions. For this study we set the population at 20, $n_{re} = 10$ and $n_{ed} = 20$ and use the four variable Rastrigin function as the objective function. The results are given in Table 6.6.

These results do not show a consistent picture since the best results are given by 20 and 70 and when $n_c = 40$ the result is worse than both of the results at 10 and 70. There are many factors interacting here and together with random effects and consequently no firm conclusions can be drawn, except the need for further study.

We now provide some graphical illustrations of the behavior of the BFOA algorithm as iterations proceed in the search for the global minimum. This is difficult to provide

Table 6.6 Shows the affect of the number of chemo-taxis steps on the performance of the algorithm

n_c	Average	Best	Worst	St Dev
10	2.1291	0.1260	5.4211	1.3912
20	1.1259	0.2127	1.7916	0.4058
40	1.3593	0.4846	2.5080	0.5733
70	1.1777	0.1447	1.9285	0.4895

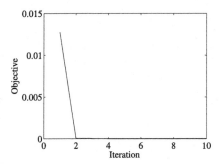

Figure 6.1 Plot showing convergence of to the global solution using BFOA for the single minimum Rosenbrock function.

since the algorithm does not have one simple set of main iterations or generations but every main iteration includes a number of nested loops. Consequently the main iterations only provide a broad view of how the search is proceeding.

Figure 6.1 shows that for the Rosenbrock function which has only one minimum, convergence is rapid and direct to the correct minimum of 0 at [1, 1] there are no other minima to trap or delay the search process. It should be noted that the graph shows the progress for the major iterations only. Consequently in terms of the main iterations the near optimum is reached very rapidly and the value shown on the graph is already very close to the minimum value of the objective function. A population of 30 bacteria is used.

Figure 6.2 shows the minimization of the Rastrigin function and in terms of the major iterations it is close to the global minimum after only a few of the iterations. However there is some improvement in the objective function even in the later stages of convergence and this occurs after some iterations where no improvement is achieved.

In Figure 6.3 the progress of minimization is shown for the four variable version of Rastrigins function. The higher dimension function presents a much more challenging problem for the BFOA and the bacterial population is relatively small. This is reflected in the value obtained for the global minimum which is significantly greater than the exact objective function minimum at zero. Figure 6.4 shows the effect of increasing the size of the bacteria population.

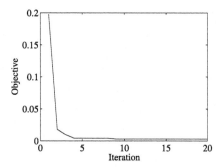

Figure 6.2 Plot showing convergence for the Rastrigin function with two variables.

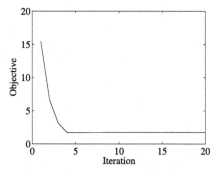

Figure 6.3 Plot showing convergence for the Rastrigin function with four variables and bacterial population size 20.

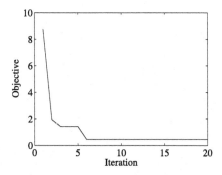

Figure 6.4 Plot showing convergence for the Rastrigin function with four variables and a bacterial population size increased to 40.

In Figure 6.4 we see that with an increased population of 40 an improved minimum value is obtained for the global minimum. Of course the population size could be increased even further and researchers often use large population sizes for higher

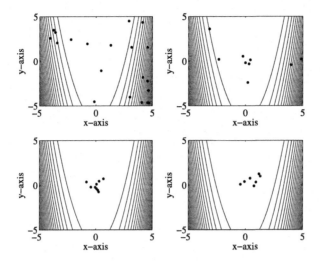

Figure 6.5 Plot showing convergence of the bacterial population for Rosenbrock's function at 0, 1, 5 and 20 major iterations.

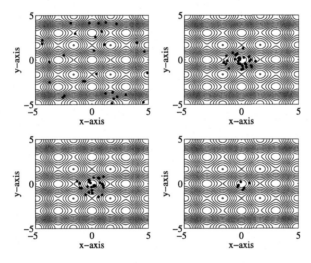

Figure 6.6 Plot showing convergence of the bacterial population for the egg-crate function at 0, 1, 5 and 20 major iterations.

dimension or challenging problems. However this should be done with care since the process of solution becomes much slower as the population size is increased.

Figures 6.5 and 6.6 show the behavior of the bacteria population as a whole during the solution process for selected functions.

Figure 6.5 shows the convergence of the bacterial population for Rosenbrock's function. Due to fast convergence in the two variable case and the large number of intermediate iterations only the first few major iterations and 10th iteration shown.

Figure 6.5 shows that the bacterial population initially placed at random points in the region where the Rosenbrock function is defined and moving towards to the minimum by the 20th iteration. There are 30 bacteria but many become superimposed at the later stages.

Figure 6.6 shows that the bacterial population initially placed at random points in the region where the egg-crate function is defined and converging to the minimum by the 20th iteration. There are 30 bacteria but many become superimposed at the later stages. We note that there is little improvement between the first and fifth major iteration but significant convergence to the optimum by the 20th major iteration.

6.5 THEORETICAL DEVELOPMENTS OF THE BFO ALGORITHM

In this section we return to discuss in more detail the work of Dasgupta et al. (2009) which was discussed in Section 6.3. These authors, as part of this paper, consider the theoretical basis for the BFO algorithm and in this section we will consider there analytical treatment in more detail. Note that this is only a one dimensional theoretical study.

Their argument proceeds as follows. Assuming the location of the bacteria at time t is given by $x(t)$ and the objective or fitness function value at this point is $J(x(t))$ where J is continuous and differentiable and the step C is small. We can define the velocity V of the bacteria for small increments in time producing small changes in x, by

$$V = \Delta x / \Delta t$$

In the limit as Δt tends to zero this becomes:

$$V = dx / dt$$

In the BFO algorithm the change in x is achieved by using (6.1) that is

$$x^i(j+1, k, l) = x^i(j, k, l) + C(i) \frac{u(i)}{\sqrt{u(i)^\top u(i)}}$$

In the one dimensional case the increment in x is equivalent to $C\delta$ where δ is the unit vector.

Now the difference in the objective function may be written as

$$J(x) - J(x + \Delta x)$$

However we have seen that in the swimming step the change in x is only permitted if it provides an improvement in the objective function, consequently this step can be represented by (6.11):

$$\Delta x = H\left(\frac{J(x) - J(x + \Delta x)}{\Delta t}\right) C\delta\Delta t \tag{6.11}$$

Here H is the Heaviside function which is defined as:

$$H(Q) = \begin{cases} 1 & \text{if } Q > 0 \\ 0 & \text{if } Q \leq 0 \end{cases} \tag{6.12}$$

This ensures by virtue of (6.11) and (6.12) that the increment for x only occurs if $J(x + \Delta x) < J(x)$ that is an improvement will be achieved in the objective function or:

$$J(x) - J(x + \Delta x) > 0$$

This implies when minimizing there is an improvement in the J value for this change in x. Now (6.11) can be rewritten on taking the limit as Δt tends to zero as

$$V = H(-GV)C\delta \tag{6.13}$$

where G is the gradient of J and V is the velocity. The Dasgupta et al. (2009) note this will mean that the bacteria will move in the direction of the negative gradient like the standard steepest descent algorithm. To avoid the discontinuity in the Heaviside function the authors replace this by the continuous logistic function:

$$CL(Q) = 1/(1 + e^{-kQ}) \tag{6.14}$$

For large values of k this has the same effect as the Heaviside function, $H(x)$, but is continuous see Figure 6.7.

Thus (6.14) becomes

$$V = \frac{1}{1 + e^{kGV}} C\delta \tag{6.15}$$

However when kGV is small, the exponential function can be approximated by using the first two terms in its expansion thus we have:

$$V \approx \frac{1}{2 + kGV} C\delta \approx \left(1 - \frac{kGV}{2}\right) C\delta/2 \tag{6.16}$$

The next step is achieved using the binomial expansion and assuming $kGV/2$ is small. We can solve for V and after further algebraic manipulation and approximation

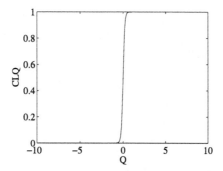

Figure 6.7 Plot of the continuous logistic function for $k = 10$, notice the rapid change from zero to 1.

provides the result:

$$V \approx \frac{kC^2}{8} G + \frac{C\delta}{2} \qquad (6.17)$$

An approximation to the velocity on a unit time interval is given by $V = (x(n) - x(n-1))/1$. Thus by virtue of (6.17) we have the iterative relation:

$$x(n) = x(n-1) - \frac{kC^2}{8} G(n-1) + \frac{C\delta(n-1)}{2}$$

where n stands for the specific time interval and $n-1$ for the one before it. Thus for specific functions Dasgupta et al. have an iterative process for finding successive value of x. They also derive the inequality (6.18) which determines a range of values for the step size C as:

$$0 \le C \le \frac{2}{k|G|}(1 + \sqrt{5}) \qquad (6.18)$$

where G is the gradient of the function being optimized, thus we see from this formula that if G is large then the upper bound on C is smaller and steps will themselves be smaller and if G is small then lager steps are permissible. This correctly reflects the nature of the surface being optimized. These theoretical considerations provide a useful insight into the behavior of the BFO algorithm.

6.6 SOME APPLICATIONS OF THE BACTERIAL FORAGING OPTIMIZATION

Bacterial Foraging Optimization is sometimes used with other optimization methods for specific practical optimization problems. A specific application where BFOA is hybridized with the Ant Colony Optimization algorithm is to Job Scheduling, a well established class of demanding problems. Here processing time of jobs must be optimized subject to the constraints that only one machine can be allotted to one job and one job to one machine. This application was described by Narendar and Annadha (2012).

A further application of BFO is described by Hezer and Kara (2012). This relates to optimal vehicle routing with simultaneous delivery and pickup. Like many problems relating to optimal routing this is an NP hard problem. This study by Hezer and Kara shows that BFO gives promising results.

When designing aerial or antenna arrays the engineer seeks to optimize the performance of the array. The key features of the array's performance are described in terms of its directivity, input impedance, beam width and side lobe level. This is a challenging optimization problem and Mangaraj et al. (2011) have used the Bacterial Foraging algorithm to optimize the design of Yagi-Uda arrays. The same authors, Mangaraj et al. (2013) revisit the design of the well established log periodic dipole array and they apply the BFO to optimize the design of the array for the complete ultra high frequency (UHF) spectrum over.

Cleaning up, or de-noising an image is an important process in many areas. Yaduwanshi and Sidhu (2013) have described how the BFO algorithm may be applied to the difficult problem of removing Gaussian noise, sometimes called 'salt and pepper' noise, from an image. This is very important in dealing with medical images of internal organs obtained by CT and MRI scans, for example, of the pancreas. Yaduwanshi and Sidhu report that the restored image shows considerable improvement in quality, measured in terms of the peak signal to noise ratio and mean square error.

6.7 SUMMARY

In this chapter we have described the BFO algorithm and how it is based on the behavior of bacterial populations using random tumbling and swimming to converge on high nutritional areas. In addition we describe some selected modifications to the basic algorithm that have been reported as producing significant improvements. We have also provided a range of numerical studies of the performance of the algorithm using different parameter values with the aim of highlighting any critical parameters effecting its performance. These studies are made on relatively small problems so that they can be easily replicated by the reader and are not meant to be original research studies. The performance of the algorithm is also illustrated graphically wherever suitable.

6.8 PROBLEMS

6.1   This exercise uses the function designed to encourage swarming of bacteria defined in (6.2). Take values for the constants as $d_a = 0.1$, $w_a = 0.2$, $h_r = 0.1$ and

$w_r = 3$. Assume the problem being optimized has two variables and the population has four bacteria. Assume these are located at the points for $i = 1, 2, 3, 4$ which are [1, 1], [2, −1], [2, 0] and [2, 3]. In addition the current point for which the computation is performed is $\mathbf{x} = [1.5, 2.5]$. Hence calculate the value of J_{cc}. If the objective function is $f(\mathbf{x}) = 2x_1^2 + (x_2^2 − 2)^2$ calculate the objective function at the point $\mathbf{x} = [1.5, 2.5]$, hence find an expression for the augmented objective function, J_{ar} for (6.3) in terms of M.

6.2 The equation which implements tumbling in Passino et al. algorithm is defined by (6.1). Assuming the current value of the $\mathbf{x}$ vector giving the bacteria location is [1, 1.5] use this equation to calculate a new vector using (6.1). You may assume $\mathbf{C}$ is the vector [0.1, 0.1]. The values of the vector $\mathbf{u}$ are uniform random values in the range [−1, 1] select these arbitrarily. Repeat this process for the vector iteratively two further times. This corresponds to the swimming process when an improvement has been achieved in the objective function.

6.3 Assuming bacteria locations are as given in Problem 6.1 and the same objective function is employed, use (6.4) to calculate new step values $\mathbf{C}$ at the points: [0.5, 1], [0.2, 1.5], [0.05, 1.6] and [0.001, 1.999], take $\lambda = 10$. Using equation (6.5) and assuming the minimum of the objective function is 0, find a new set of values for $\mathbf{C}$ using the same points. What do you note about the values of $\mathbf{C}$ you have produced?

CHAPTER 7

Artificial Bee and Ant Colony Optimization

7.1 INTRODUCTION

Both the Ant Colony Optimization algorithm and the Artificial Bee Colony algorithm involve modeling of the co-operation between insects to obtain an optimum outcome for the colony as a whole. The Ant Colony optimization method has been particularly effective in solving the traveling sales person problem (TSP), a very difficult (formally an NP hard) problem. Consequently we will discuss the TSP problem and its solution using Ant Colony Optimization (ACO). Ants convey information between members of the colony using pheromones. The ABC algorithm also involves co-operation and the bee dance or waggle dance is used to convey important information to the other bees of the hive by returning bees regarding the location of nectar. This method is directly applicable to the solution of nonlinear optimization problems, rather than the TSP problem.

7.2 THE ARTIFICIAL BEE COLONY ALGORITHM (ABC)

This algorithm was introduced by Karaboga (2005) and developed in Karaboga and Basturk (2007). It involves a cooperative swarming method inspired by the behavior of honey bees and the method has had considerable success in its application to the solution of difficult global optimization problems. In the ABC algorithm the locations of the food source, nectar, represent possible solutions of the optimization problem and the amount of nectar at these positions the quality or fitness of the solution. The bees in the colony are the key to locating these positions instinctively they achieve this in the most efficient or optimal way. To do this the bee colony is divided into three groups: onlookers, employed and scouts. Half the bees are employed and the other half are onlooker bees. Scout bees are introduced as the process evolves. For every nectar food source there is an employed bee. An employed bee can search the local area for the nectar sources. It visits a source until the source is exhausted. Then that bee is designated as a scout bee and performs a random search of the whole area. Once a scout bee finds a new source of nectar, its location is memorized by the individual bee and the new location replaces the exhausted one. After each employed bee has completed its search cycle it exchanges this information with the onlooker bees. In a real bee colony this is achieved by a bee "waggle dance", remarkably this formalized dance can convey to the onlookers the direction, distance and quality of the nectar source. The onlooker bees analyze all the information on position and nectar amounts and decide on the most fruitful areas. They then fly to the most promising nectar sources. These concepts can be utilized in the development of an optimization algorithm. We now provide a more formal description of this process.

Initially a set of candidate solutions are generated randomly within the solution space each one associated with an employed bee.

At each iteration the employed bee tries to discover a new possible solution by searching its neighborhood this can be formulated using the equation:

$$v_{ij} = x_{ij} + r_u(x_{ij} - x_{kj}) \tag{7.1}$$

where r_u is a uniformly distributed random number in the range $[-1, 1]$ and is a different random number for every value of i and j, k and j are randomly selected from the set of indices of employed bees and the dimension indexes of the problem, respectively. The variable x_{ij} is the current solution and v_{ij} the candidate improved solution. The subscript i denotes the employed bee and j the particular dimension. The new value is checked for improvement using the fitness function if there is an improvement this employed bee moves to this new source of nectar.

Once this process is complete the information from each of the employed bees is presented to the onlooker bees who make an overall judgment of these nectar sources. They then choose one with a probability p_i based on the relative value of the particular source compared with all the other sources. This is implemented by computing the values, called fitness values from the original objective function of the problem $f(x)$, using (7.2).

$$fit_i = \begin{cases} \frac{1}{(1+f(x_i))} & \text{if } f(x_i) \geq 0 \\ 1 + \text{abs}(f(x_i)) & \text{otherwise} \end{cases} \tag{7.2}$$

and then the values of the probabilities are calculated from

$$p_i = \frac{fit_i}{\sum_{i=1}^{m} fit_i} \tag{7.3}$$

where fit_i is the fitness of solution i, the summation is for all employed bees. These values of p_i are in the range $[0, 1]$. An alternative method suggested for calculating the probability values uses the formula:

$$p_i = 0.9 \frac{fit_i}{fit_{best}} + 0.1 \tag{7.4}$$

where fit_{best} is the best fitness value. Dependent on the probability values new points are generated using (7.1). An important feature of the algorithm is to abandon solutions that are not fruitful in that they have not provided an improvement in the objective function value after a number of cycles. This is implemented by setting a limit parameter which is preset by the user. If the number of times there has been no improvement at that location exceeds this limit value this location is abandoned and replaced with new randomly generated solution within the solution space using (7.5). This can be considered as the exploratory stage and models the role of the scout bee:

$$x_{i,j} = r_{min,j} + r_u(r_{max,j} - r_{min,j}) \tag{7.5}$$

where r_u is a uniform random number in the range [0, 1]. The value of limit should be carefully considered and its setting should relate to the difficulty of the particular problem; e.g. it should increase with the size or complexity of the problem. This process is repeated at each cycle until the required number of iterations have been performed. We are now in a position to provide a pseudo code formulation of the method on which its implementation can be based. This is shown in **Algorithm 6**.

This completes the general description of the method and many studies have reported good comparative results for this algorithm. Various suggestions have been made by different researchers about the values of the given constants and other features of the ABC algorithm and we shall discuss a selection of modifications of the basic algorithm in the next section.

7.3 MODIFICATIONS OF THE ARTIFICIAL BEE COLONY (ABC) ALGORITHM

The first modification we shall consider for the ABC algorithm is that suggested by Li et al. (2014). Li et al. point out that there algorithm is a further development of work published in an earlier paper (Li et al., 2014) and they begin their discussion of improvements to the original algorithm with a description of this work which we will now describe.

The algorithm introduced in 2014 uses the vector **trial** which keeps count of the number of times an employed bee has failed to find an improved location for a nectar resource. The values of the elements of this vector **trial** are used to guide the process of exploration and exploitation that are found in the original ABC algorithm and may be described as follows:

$$x_{ij}^* = x_{ij} + r_u(x_{ij} - x_{kj})\gamma_i \tag{7.6}$$

where r_u is a uniform random number in the range -1 to 1 and is different for each value of i and j, and γ_i is defined by the equation

$$\gamma_i = \exp\left(-(trial_i - 1)\frac{\log_e 10}{(n_{var} - 1)}\right)$$

Algorithm 6 Basic ABC.

Set cycle =1; Set constant parameters: maxcycle, N_s, limit, where N_s is the number of nectar sources.

Generate initial bee population randomly within the defined solution space for the specific problem

while *cycle < maxcycle* **do**

    Calculate the new locations $v_{i,j}$ for the employed bees using

    for $i = 1$ **to** N_s **do**

        for $j = 1$ **to** *nv* **do**

            Compute $v_{i,j}$ from (7.1)

        end for

    Where k and j are randomly selected integers in the appropriate range. Note that $k \neq i$.

    Evaluate the fitness of the new sources based on the objective function

    if *fit*$(\mathbf{v}_i)$ > *fit*$(\mathbf{x}_i)$ **then**

        $\mathbf{v}_i$ replaces $\mathbf{x}_i$.

    end if

    end for

    for $i = 1$ **to** N_s **do**

        Compute p_i from (7.3) where *fit*$_i$ is given by (7.2)

    end for

    The onlooker bees select new solutions v_i based on the probabilities p_i and evaluates them using the fitness function and replace the old solutions with a new one if an improvement is achieved. If for a particular source s no improvement is achieved keep a record of that by setting *trial*(s) = *trial*$(s) + 1$. Otherwise the value of *trial*(s) is set to zero.

    if Fitness of a nectar source has not improved for greater than *limit* iterations **then**

        abandon it and replace it by a randomly selected source location in the solution space using (7.5) this models the role of the scout in exploring new areas.

    end if

    Store the best solution so far.

    cycle = cycle + 1

end while

Now the value of each *trial* lies between 1 and n_{var}, where n_{var} is the number variables in the problem. This means that the value of γ_i varies between 0.1 and 1 because for each employed bee i, *trial*$_i$ may vary between n_{var} and 1, since if *trial*$_i = 1$ then $\gamma_i = 1$ and if *trial*$_i = n_{var}$ then $\gamma_i = \exp(-\log_e 10) = 1/\exp(\log_e 10) = 0.1$. This means the random element of exploration is reduced if failures are becoming high, thus intensifying the search in the local region before it is abandoned. The next development described by Li et al. which constitutes their new algorithm also introduces a parameter that varies through the generations but this is based on more complex use of the trial vector. This introduces the update equation:

$$x_{ij}^* = x_{mj} + r_u(x_{kj} - x_{ij})\mu_i \tag{7.7}$$

where r_u is a uniformly distributed random number in the range [0, 1]. (7.7) is similar to (7.6) but uses the index m and the new variable factor μ_i. The j index is as usual the index of the specific dimension and consequently $j = 1, 2, ..., n_{var}$, k and m are specific members of the bee population and k is not equal to i. The factor μ_i is defined by (7.8):

$$\mu_i = trial_i/(trial_i + trial_k) \tag{7.8}$$

As before trial is confined in the range $[1, n_{var}]$ and provides the number of times the source location has failed to produce an improvement in the specified food source quality and is updated accordingly. Thus the effect of the random term is modified by the factor μ_i according to information about the current quality of the sources. Li et al. point out that the value of μ_i is confined by the inequality:

$$\frac{1}{n_{var} + 1} \le \mu_i \le \frac{n_{var}}{n_{var} + 1} \tag{7.9}$$

Thus the range of values for μ_i lies within the range [0, 1], since if n_{var}, the number of variables, becomes large the end points of the range tend to 0 and 1 and if small still lies within this range. For example if $n_{var} = 2$ then $1/3 \le \mu_i \le 2/3$. To consider the effect of the μ_i factor we note that if $trial_i$ is small it means that the source is currently good; if $trial_i$ is large then source is not efficient. This value effects μ_i and if $trial_i$ is small so is μ_i. This means less emphasis on exploration. However, if $trial_i$ is large so is μ_i and this encourages exploration. Li et al. hope this feature provides a more nuanced balance between exploitation and exploration. Li et al. also introduce the factor to the onlooker bee stage of the algorithm by using (7.10).

$$Y_{0,j}^* = x_{0j} + r_u(x_{kj} - x_{0j})\mu_i \tag{7.10}$$

where r_u is a uniform random number in the range $[-1\ 1]$ and is different for each j. Here the local investigation of the solution space is again modified by using the factor μ_i. We note that at any time if the value of any trial factor exceeds n_{var} then it is reset to n_{var}. Towards the end of each cycle of the algorithm the average of all the trial values associated with each bee are computed. Then this average trial value is compared with $0.9n_{var}$ if it is smaller then the employed bees are recycled using (7.5) since they are

---

Algorithm 7 Modified ABC by Li et al.

Set initial values for all parameters.
Generate randomly the initial population within the solution space.
set all trial values to 1.
for $gen = 1$ **to** *maxgen* **do**
    for $i = 1$ **to** N_s **do**
        for $j = 1$ **to** n_{var} **do**
            Compute $x_{i,j}*$ from (7.7)
        end for
    end for
    if improvement then accept new point and update trial value **then**
        $trial(i) = 1$
    else
        $trial(i) = trial(i) + 1$
    end if
end for
Calculate probabilities for employed bees using (7.3) or (7.4).
set $i_b = 1$
while $i_b < N_s$ **do**
    mutate for one element using (7.10).
    accept new value if improved and update trial values.
    $i_b = i_b + 1$;
end while
for $i = 1$ **to** N_s **do**
    if $trial(i) > n_{var}$ **then**
        $trial(i) = n_{var}$
    end if
end for
if average trial $> 0.9n_{var}$ **then**
    reinitialize 0.9 of employed bees using (7.5)
    record best solution
end if

---

considered inefficient. The pseudo-code for this given by the authors Li et al. takes the form of **Algorithm 7**.

Li et al. have applied there modified algorithm to a vehicle path optimizing problem and the results show significant improvement particularly for high dimensions on the original ABC algorithm. It would be interesting to see how this modification worked in relation to the standard test set of non-linear optimization problems.

The final Modification to the ABC algorithm we will consider is that proposed by Gao and Liu (2011). These authors introduce two major changes to the algorithm. Firstly instead of the usual random generation of the initial population, they propose to generate the initial population based on chaotic systems and opposition based learning. Their justification for this is that this process can provide improved initial values which should result in rapider convergence. The second modification of Gao and Liu is to introduce a search procedure based on Differential Evolution which we have discussed in Chapter 2. We provide more details for these interesting suggestions. The chaotic generation of points in the solution space is achieved using the sine function and is given in (7.11),

$$c_{k+1,j} = \mu c_{k,j}(1 - c_{k,j}), \text{ for } k = 0, 1, 2, ..., K \tag{7.11}$$

for all j. Initially the $c_{0,j}$ are randomly selected in [0, 1], the value K is the user set number of chaotic iterations and j denotes the index of the dimension of the problem. Having generated the value of c_{kj} we can generate values that lie within the solution space using the equation:

$$x_{ij} = r_{min,j} + c_{k,j}(r_{max,j} - r_{min,j})$$

where $i = 1, 2, ..., n_s$ is the index of the members of the population, $j = 1, .., n_{var}$ the index of the variables and $r_{min,j}$ and $r_{max,j}$ are the lower and upper bounds of the problem solution space for each variable. In the second stage of this process opposition based learning is used, this procedure is described in Rahnamayan et al. (2008). A new set of points are then generated using (7.12)

$$opx_{ij} = r_{min,j} + r_{max,j} - x_{ij} \tag{7.12}$$

Finally the best n_s individuals of the $opx_{i,j}$ and $x_{i,j}$ are selected using the objective function to constitute the new population. The next modification which seeks to improve the exploitation element of the ABC algorithm is based on differential evolution and takes the form:

$$v_{i,j} = x_{best,j} + r_u(x_{r1,j} - x_{r2,j}) \tag{7.13}$$

where r_u is a uniformly selected random number in the range $[-1, 1]$. The indices r_1 and r_2 are different values chosen randomly from the indices 1, 2, ..., n_s. An important feature of this amendment is that $x_{best,j}$ is the current best member of the bee population. This equation drives the exploitation phase of the algorithm around the current best value. Gao and Liu have given a useful outline of their modified algorithm.

Gao and Liu have carried extensive numerical tests on the performance of their algorithm on a wide range of test problems and have reported significant improvements on the performance of their algorithm when compared with the original ABC algorithm. This completes the description of our selection of modifications of the ABC algorithm.

Table 7.1 Minimization of the function RAS6 using various numbers of iterations

Iterations	Mean	Best	Worst	St Dev
100	1.5358	0.0030	2.9024	0.8192
200	0.0174	1.4646×10^{-4}	0.1573	0.0441
500	3.5527×10^{-16}	0	7.1054×10^{-15}	$1.58884646 \times 10^{-15}$

Table 7.2 Minimization of the function RAS6 using various numbers of food sources

food Sources	Mean	Best	Worst	St Dev
10	0.3161	$6.09464646 \times 10^{-4}$	1.0924	0.4333
20	0.0322	4.4813×10^{-5}	0.1751	0.0548
30	0.0047	1.5159×10^{-5}	0.0376	0.0092

Table 7.3 Minimization of the function RAS6 using different probability formulae

Formula	Mean	Best	Worst	St Dev
Using (7.3)	0.0089	9.8661×10^{-6}	0.0780	0.0177
Using (7.4)	0.2235	6.2507×10^{-4}	1.0634	0.3544

7.4 SELECTED NUMERICAL STUDIES OF THE PERFORMANCE OF THE ABC ALGORITHM

In this section we will study the behavior of the ABC algorithm when tested on standard test problems. In each case, ten or twenty runs of the algorithm are performed for the numerical studies. However because of the random nature of the process caution should be exercised before coming to conclusions about the results. Many runs of the algorithm are used to average out some of the effects of this randomness in each of the studies. The first test we undertake is to study the effect of the number of iterations used. The exercise uses the Rastrigin function with 6 variables, RAS6, 30 nectar sources a limit value of 100 before the scout bees take part. The results are provided in Table 7.1.

As expected, Table 7.1 shows that increasing number of iterations improves the performance of the algorithm minimizing Rastrigins function with six variables.

In Table 7.2 we show the effect of changing the number of food sources available to the bees. We use 200 iterations for these experiments a limit of 100 and again Rastrigin's function with six variables.

Table 7.2 shows that the performance when there are only 10 food sources is not strong since the result is distant from the true global minimum of zero. However the result markedly improves with an increase in the number of food sources to 20 and when the number of food source is 30 the best solution is found.

In Table 7.3 we consider the effect of choosing alternative formula for the probability calculation, which have been suggested, at the onlooker bee stage these are described

Table 7.4 Minimization of Rastrigin's functions with different numbers of variables

Function	Mean	Best	Worst	St Dev
RAS6	0	0	0	0
RAS8	1.6045×10^{-8}	1.1923×10^{-11}	0.0780	5.1134×10^{-4}
RAS10	9.3491×10^{-4}	9.9657×10^{-8}	0.0018	4.0658×10^{-4}
RAS15	0.9401	0.0322	2.0076	0.5565

as form1 and form2. Where the formula form1 is given by (7.3) and form2 is given by equation (7.4) in Section 7.2 of this chapter. The experiments minimize RAS6 and are carried out using 500 iterations and the number of food sources is 30.

We see that Table 7.3 shows a significant improvement in using (7.4) rather than using (7.3). However much further work on a range of problems would be required to establish this result.

In the last table, Table 7.4 we see the effect of increasing dimensionality on the difficulty of solving the problem by studying the algorithms performance on the six variable Rastrigin function (RAS6) and the eight variable Rastrigin function (RAS8) and ten variable Rastrigin function (RAS10) and finally the 15 variable Rastrigin function (RAS15). Five hundred iterations are used with 30 food sources for each dimension.

Table 7.4 shows that the results become less accurate with the increasing dimension of the problem, in particular the 15 variable problem has a relatively poor result and this does demonstrate how the increase in the dimensions of the problem is a very influential factor on the performance of the algorithm.

We now give some graphical insight into the performance of the ABC algorithm by drawing contour graphs of the function and displaying the changing position of the bees and food sources as the algorithm progresses. Figure 7.1 shows the contour plots of the Rosebrock function with two variables.

Figure 7.1 shows contour graphs at 0, 10, 50 and 1000 iterations. The trial solutions have clustered to the same point after 1000 iterations. There is only one global optimum for this problem and it lies in a very shallow valley. The algorithm appears to locating points along a contour of the function here the objective function values will be close in value.

We now consider the two variable Styblinski-Tang function. Figure 7.2 shows the contour plots for the initial locations and at 10 iterations, 50 iterations and after 200 iterations. In this case convergence is very rapid to the solution at $(-2.9035, -2.9035)$, the contour graphs show the situation at 0, 10, 50 and 200 iterations.

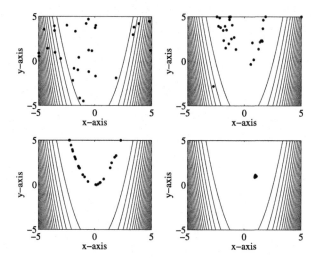

Figure 7.1 Optimization of Rosenbrock's function showing clustering along the optimum level contour and then convergence to the solution at the point [1, 1] with the function value zero. At the initial locations and after 10, 50 and 1000 iterations.

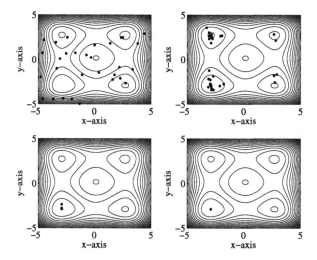

Figure 7.2 Optimizing the Styblinski-Tang showing initial nectar locations and nectar locations after 10 iterations, 50 iterations and 200 iterations.

The final example shows the contours of the egg-crate function and the state of the convergence at 0, 10, 50 and 200 iterations in Figure 7.3. This function has very many minima quite closely packed together. The swarm approaches the global minimum very rapidly and the points become practically coincident after only 200 iterations of the algorithm.

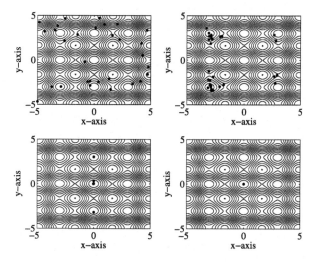

Figure 7.3 Optimization of the egg-crate function. Rapid convergence after some initial exploration of the region. Rapid convergence occurs to the true minimum at [0, 0]. The contour graphs show the initial position and after 10, 50 and 200 iterations.

7.5 SOME APPLICATIONS OF ARTIFICIAL BEE COLONY OPTIMIZATION

Hossain and El-shafie (2013) use the ABC algorithm to optimize the reservoir release per month at the Aswan high dam. The algorithm was used to solve the problem based on actual historical inflow data and it succeeded in meeting the demand over a specific period.

Bolaji et al. (2013) provides a survey of modifications to the ABC algorithm and indicate a number of applications of the algorithm. They include: stock price forecasting, artificial neural networks, image processing, electric load forecasting and flow job scheduling.

We now discuss in more detail applying the ABC algorithm specifically to the important problem of optimal financial portfolio selection. The paper of Bacanin et al. (2014) describes how the authors have applied the ABC algorithm to minimizing risk in investment portfolio selection which is major financial problem.

Many optimization problems arise in the world of finance. An important one of these is one which arises in establishing a profitable portfolio of investments. An approach to this would be to maximize the return from a selected group of investments, however Bacanin et al. (2014) discuss the alternative approach is to minimize the risk of the portfolio of investments by selecting the investments according to some criteria. This is an established problem and mathematical formulation has been provided by Markowitz (1952). Later in 1990, Markowitz, Miller and Sharpe won the Nobel prize for work in the theory of financial economics and corporate finance. This formulation of the minimum risk approach may be expressed in the form

$$\text{minimize } \sum_{i=1}^{n} \sum_{j=1}^{n} w_i w_j \, \text{cov}(R_i R_j)$$

$$\text{subject to } \sum_{i=1}^{n} w_i R_i \geq R$$

$$\sum_{i=1}^{n} w_i = 1$$

$$w_i \geq 0, \text{ for all } i = 1, 2, ..., n$$

Here w_i is the weight or proportion of the investment in the portfolio, n is the number of individual investments and R_i is the mean return on investment i. The value R represents the required minimum return for the selected portfolio. The coefficients $\text{cov}(R_i R_j)$ are the correlation coefficients between the returns i and j. In this formulation the aim is to find the values of w_i which are the proportions of the investments in the portfolio which minimize the risk or variance of the portfolio subject to constraints on the investment return requirement. The sum of the proportions w_i must be 1 and each w_i must be positive. Although the problem could be reformulated to maximize returns. Mathematically this problem is classified as a constrained quadratic programming problem.

Bacanin et al. note that the constraints on investment returns can be reflected into the objective function by introducing the parameter λ, this parameter must be positive and lies in the interval [0, 1]. The new objective function now provides a balance between risk and return and Bacanin et al. indicate that this as may be used as an alternative formulation for the original problem. The problem may be solved using a range of methods the classical approach being to use Lagrange multipliers. However the methods can be included in the objective function using penalty function methods. Note that the general solution of constrained non-linear optimization problems is discussed in Chapter 9. Bacanin et al. note that in practical applications many important additional factors can affect the formulation of this problem. For example the proportion of the investments may not be continuously variable, there may be legal restraints on the simultaneous holding of certain assets, and there may also be economic constraints on the problem. Consequently they reformulate the problem taking into account some of these issues which includes additional constraints. However Nature Inspired optimization methods could clearly be used for this type of problem and its variants and these can supply the global optimum for the problem.

Bacanin et al. then describe the ABC algorithm and apply it to historical data providing the performance of five stocks from 2007 to 2011. Bacanin et al. apply the basic model and report the results for the weights found by the ABC algorithm and state that it has the potential to solve this type of problem. Another application similar to this one is given by Chen (2014). However this application is to uncertain portfolio selection. The returns are based on expert prediction rather than historical data. An interesting feature of this application is that the problem is formulated as a three objective non-linear programming problem subject to constraints. Where the three competing objectives are to minimize risk, to maximize return and to maximize the diversity of the portfolio. Interestingly the diversity objective uses the standard entropy function. The solution of multi-objective function problems is discussed in Chapter 9 in this text and a range of methods of solution for the problem are given. Chen applies the ABC method to specific numerical examples in portfolio selection. He indicates the results show the ABC algorithm is effective for solving the portfolio selection problem.

7.6 DESCRIPTION OF THE ANT COLONY OPTIMIZATION ALGORITHMS (ACO)

The term Ant Colony Optimization (ACO) is a collective term which is used to describe a group of algorithms which use a cooperative swarming method based on modeling the behavior of ant colonies when foraging for food. This group of algorithms was introduced by Dorigo et al. (1996) and, Dorigo and Gambardella (1997) when they used the term Ant System (AS) to describe one of the first versions of the ACO methods and this is the one we shall describe in detail. Like the ABC algorithm information is passed from some members of the colony to others. In the ACO group of algorithms this is achieved by placing pheromones rather than a waggle dance as in the ABC algorithm.

Whilst individual ants can only do very simple tasks, a colony working together can exhibit intelligent behavior. As an ant walks it deposits a chemical called a pheromone that encourages other ants to follow it. The pheromone evaporates over time and this encourages the ants to explore other routes in their search for food. The greater the number of ants that follow a particular path, the stronger the trail of pheromone. The density of pheromone deposited also increases more rapidly on shorter paths and this encourages more ants to take these shorter paths.

From the natural observed behavior of a colony of ants, Goss et al. (1989) developed a formula which gave the probability that an ant at a junction of two branches, would select a particular branch of the two. The formula takes the form:

$$p_1 = \frac{(m_1 + k)^h}{(m_1 + k)^h + (m_2 + k)^h}$$

Here p_1 is the probability of taking the first branch and k and h are constants fitted to the recorded behavior of the ants, Goss et al. found $k = 20$ and $h = 2$ provided good results for the set of observations they made. The values m_1 and m_2 provide the number

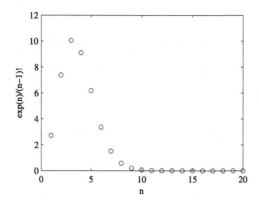

Figure 7.4 Illustrates the comparison between $(n-1)!$ and e^n.

of ants in the colony who have so far chosen the first and second branch respectively. Clearly the probability of taking the second branch p_2 is $1 - p_1$ when there are only two such options. Dorigo et al. utilized and extended this concept in generating their ant colony algorithm.

These principles may be usefully applied to optimization problems in particular they can be easily modified to find the solution of particularly difficult problem known as the symmetrical traveling salesman problem or TSP which seeks to find the optimum route for a salesman who must visit a number of cities and return to the starting point of his journey by a route which will minimize the distance the salesman travels in visiting all the cities once. It is one of a number of difficult combinatorial problems which present major problems for the effective solution, e.g. the Knapsack problem and the job assignment problem.

The TSP is simple to describe but it is very difficult to solve. To solve this problem by enumeration of alternative routes for n cities involves the evaluation of $(n-1)!/2$ distinct tours (division by two since back and forth along the same route does not have a different cost). The value of $(n-1)!/2$ increases rapidly with n, hence the difficulty of the problem. For example for a five city example the method of enumeration would involve 4! or 24 possible tours but 20 cities would involve $19! = 1.2164510004 \times 10^{17}$ possible tours. It is estimated that using the direct enumeration method for 20 cities could take over 190,000 years. Despite this challenge the size of problem that can be solved is increasing rapidly for example the 1980s saw the solution of problems with 318 cities while in 2006 a TSP problem with 85900 cities was solved. This is due to the steady development of better and better algorithms. Figure 7.4 illustrates how the factorial function can grow much faster than the exponential function thus the ratio $\exp(n)/(n-1)!$ rapidly tends to zero with increasing n due to the factorial function value dominating the exponential function value.

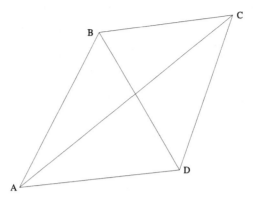

Figure 7.5 TSP problem with four towns.

The TSP problem is formally classified as NP hard which highlights its exceptional difficulty. The term NP designates the term Non-deterministic Polynomial and refers to the time taken to solve the problem. It is expected, but not proved, that no polynomial time algorithms exist to solve this type of problem. Consequently there is an ongoing search for improved algorithms that are far more sophisticated than the direct enumeration of the possible solutions which as we have seen is not a practical method for even relatively small problems. The problem may be written as a constrained minimization problem in variables having the values zero or one, for large problems this is not a practical formulation. To adapt the Ant System approach to the TSP problem we must impose specific requirements on the individual ant consistent with the behavior of a colony of ants.

1. It must visit each city exactly once;
2. A distant city has less chance of being chosen (because the more distant the city, the less the visibility);
3. The more intense the pheromone trail laid out on an edge between two cities, the greater the probability that edge will be chosen;
4. Having completed its journey, the ant has deposited more pheromones on all the edges it traversed
5. After each iteration, trails of pheromones evaporate at a specific rate.

Each TSP problem may be represented by a connected graph and an example is shown in Figure 7.5. A, B, C and D are the locations of four towns. The distances between the various points on the graph can be represented by $d_{i,j}$ where i and j are the labels of the end points on the graph. This represents the distance between the two points i and j. The concept can be related to Cartesian coordinates of the points on the graph by being defined as the Euclidean distance between them. The line connecting

i and j is called an edge. Thus members of the ant colony will traverse the edges of the graph laying pheromones as they pass. The ants will tend to choose the shortest paths and these will be enhanced by having additional pheromones deposited and thus attracting other ants. To develop an algorithm these features must be implemented in a formal way. If we define the key features of the problem as follows: the number of towns is n_t, the total ant population as m and the number of ants in each town as n_i and $\tau_{i,j}$ as the ant pheromone intensity on each trail joining the points i and j.

Dorigo et al. used (7.14) to model the manner in which ants deposited their pheromones on each trail. Thus if $\tau_{i,j}$ is the intensity of pheromone on the edge i, j this takes the form:

$$\tau_{i,j}^{(t+n_t)} = \rho \tau_{i,j}^{(t)} + \Delta \tau_{i,j} \tag{7.14}$$

where $1 - \rho$ represents the amount of evaporation of the pheromone on the trail, the quantity $\Delta \tau_{i,j}$ represents the amount of pheromone added. So (7.14) represents both how the pheromone increases and how it decreases through natural evaporation. Note that this equation represents a complete cycle of the process in that all ants have completed their tours. Thus the use of the superscript $t + n_t$ in (7.14). Hence if $\Delta \tau_{i,j}^k$ represents the amount of pheromone deposited per length of trail on edge joining i and j by the kth ant then the amount deposited is the sum of all these deposits for each ant thus:

$$\Delta \tau_{i,j} = \sum_{k=1}^{m} \Delta \tau_{i,j}^k \tag{7.15}$$

where $\Delta \tau_{i,j}^k$ is defined by (7.16)

$$\Delta \tau_{i,j}^k = \begin{cases} Q/L_k & \text{if } k\text{th ant uses edge } i, j \text{ in its tour} \\ 0 & \text{otherwise} \end{cases} \tag{7.16}$$

where Q is a constant and L_k is the tour length of the kth ant. Dorigo et al. introduce further features to their algorithm by noting that since each town must be visited only once, a list must be kept of towns visited. This they call the tabu list since if a town has been visited it must not be visited again. One of these lists is associated with each ant, consequently for the kth ant there is an associated tabu list or vector, **tabu**$_k$.

Since an individual ant makes specific choices when presented with alternative roots there must be some way of modeling this decision process. Dorigo et al. formulate this by taking into account the visibility of the next locations v_{ij} and the intensity of the pheromone trail, τ_{ij}. They propose that the specific ant k will make a choice with probability calculated from (7.17).

$$p_{i,j}^k = \frac{\tau_{i,j}^\alpha \nu_{i,j}^\beta}{\sum\limits_{k \in S_k} \tau_{i,k}^\alpha \nu_{i,k}^\beta}$$

(7.17)

Here S_k are the set of allowed indices for the kth ant i.e. the towns not yet visited, and $\nu_{i,j} = 1/d_{i,j}$. Thus, the visibility is reciprocal of the distance between the towns i and j thus if the distance is great the visibility tends to zero. The allowed towns are the set of towns not in the particular tabu list. The powers α and β can be adjusted to alter the effects of the visibility and pheromone trail intensity. For example, if $\beta = 0$ then the visibility has no effect. If $\alpha = 0$ then the pheromone intensity will have no effect. The larger α and β the more amplified the two characteristics become. Clearly if the town is highly visible and the intensity of pheromone on the trail is high then the probability of selecting this part of the route will be high. We note also that the probability will be between 0 and 1, since $\nu_{i,j}$ and $\tau_{i,j}$ are positive and sum of all the allowed combinations will always be greater than or equal to one of them.

A broad description of the ant system algorithm is given by Dorigo et al. They state that at time zero the ants are positioned in the towns or vertices of the network that defines the problem and trail intensities are associated with each edge. All the tabu lists for each ant have their first position set to the starting town of each ant. The ants are then ready to begin their journey and do so by selecting a route to take amongst the alternatives using the probability defined by (7.17). After n_t iterations, n_t is the number of towns, when all ants will have completed a tour and their tabu lists have been filled thus for each ant the length of their journey can be computed and values of pheromone can be updated using (7.16). The best route for all the ants is saved. Based on this descriptions Dorigo et al. provide a pseudo-code for the implementation of the ant algorithm. This takes the form given in the pseudo-code **Algorithm 8**.

This completes the description of the ant colony algorithm. Dorigo et al. make the important observation that since they have experimentally found that there is a linear relation between the number of towns and the best number of ants and in fact that the number of ants is close to the number of the cities, consequently the complexity of their algorithm is, they state, of order of the number of cycles times the cube of the number of towns.

Dorigo et al. carried out a range of numerical tests on standard TSP problems and found their algorithm produced successful results. Since its introduction in the 1990's the algorithm has become established as an important development in the field. The key parameters of the algorithm are α, β, ρ and Q. Dorigo et al. indicate that $\alpha = 1$, $\beta = 0.5$, $\rho = 0.5$ and $Q = 100$ is a good parameter set.

Algorithm 8 ANTS.

Initialize all the values and time and cycle counters. Set initial values for the pheromones at some small selected positive constant c. Set additive amounts of pheromone at zero. Place the m ants selected on the network nodes. Set number cycles, $n_c = 0$. Set number of towns, n_t

Begin main cycle

for $k = 1$ **to** m **do**

    Place the initial location of each ant in their tabu list

end for

Move all the ants on to the next town. Repeat $n_t - 1$ times.

for $k = 1$ **to** m **do**

    Use (7.17) to calculate the probability that the ant k moves to town j from its current position in town i.

    As the ant is now in town j this is placed in the **tabu**$_k$

end for

for $k = 1$ **to** m **do**

    move the kth ant from its tabu list position n to 1.

    Compute the length of the tour of ant k as L_k.

    Find the smallest route i.e. $\min(L_k)$ for all $k = 1, 2, ..., m$.

    if better than the current minimum length route **then**

        this is updated to the new value.

    end if

    Now calculate the new pheromone values laid for each edge and for each ant.

    for $k = 1$ **to** m **do**

        Use (7.15) to calculate values of $\Delta\tau_{i,j}^k$ then

$$\Delta\tau_{i,j} = \Delta\tau_{i,j} + \Delta\tau_{i,j}^k$$

    end for

end for

For every edge i, j compute $\tau_{i,j}$ using (7.14) to update pheromone level.

Set $t = t + n_t$, $n_c = n_c + 1$

for every edge i, j set $\Delta\tau_{i,j} = 0$

if the maximum cycles have not be reached **then**

    clear all tabu lists and repeat the main cycle.

else

    output shortest route

end if

7.7 MODIFICATIONS OF THE ANT COLONY OPTIMIZATION (ACO) ALGORITHM

Dorigo et al. briefly describe alternatives to the original algorithm which they call the ant density and ant quantity algorithms. The only difference between these versions of the algorithm is in the way in which the pheromone on the trails is updated. For the ant quantity model (7.16) is replaced by

$$\Delta_{i,j}^k = \begin{cases} Q & \text{if } k\text{th ant uses edge } i, j \text{ in its tour} \\ 0 & \text{otherwise} \end{cases}$$

For the ant density model and

$$\Delta_{i,j}^k = \begin{cases} Q/d_{i,j} & \text{if } k\text{th ant uses edge } i, j \text{ in its tour} \\ 0 & \text{otherwise} \end{cases}$$

where $d_{i,j}$ is the distance between town i and town j.

Dorigo et al. carried out studies on the relative performance of these alternative updating methods and concluded that the original updating method given by (7.16) produced the best results. Dorigo (2007), in Scholarpedia, provides a useful outline of the relationship between the main ACO algorithms.

1. The first ACO algorithm called Ant System (AS) is the algorithm which we have described in Section 7.6;
2. Ant Colony System (ACS)
3. The MAX-MIN system

We shall now describe briefly the ACS algorithm and the MAX-MIN algorithm. The ACS algorithm proposed by Dorigo and Gambardella (1997) is the first major improvement on the original algorithm and is similar to the original algorithm. However one significant difference is that it uses a different rule in deciding whether to move from location i to location j. This depends on the selection of a uniform random variable U in the range [0, 1] and a parameter U_0 which is selected by the user. Then the route is selected using (7.18).

$$\text{if} \quad U \le U_0 \text{ then } j = \underset{j \in J_{allowed}}{\operatorname{argmax}}(\tau_{ij} v_{ij}^\beta) \tag{7.18}$$

else use rule (7.17) as in the original AS algorithm.

This is strongly directed by the pheromone element but is counteracted by increasing diversity by a local pheromone updating process performed by all ants but only to the edge most recently traversed. This is achieved by (7.19).

$$\tau_{ij} = (1 - U)\tau_{ij} + U\tau_0 \tag{7.19}$$

where U is a uniform randomly selected number in the range [0, 1] and τ_0 the initial value of the pheromone.

This process will strongly increase the diversity of the method. The final modification is to perform the pheromone update but only along the edges of the current best ant route hence:

$$\tau_{ij} = (1 - \rho)\tau_{ij} + \rho(\tau_{ij})_{best}$$

where $(\tau_{ij})_{best} = 1/L_{best}$ and L_{best} is the length of the best route.

Another important development was the introduction of the MAX–MIN Ant System (MMAS); this improvement was introduced by Stuzle and Hoos (1997). This algorithm differs in two ways from previous algorithms in that only the best ant is permitted to add pheromones and the amount of pheromone that can be added is constrained between an upper and lower bound. Specifically this means that

1. The pheromone update takes the form:
   $\tau_{i,j} = (1 - \rho)\tau_{i,j} + \Delta\tau_{i,j}^{best}$ where $\Delta\tau_{i,j}^{best} = 1/L_{best}$.
   If the best ant used edge (i, j) in its tour otherwise it is taken as zero.
2. The pheromone amount $\tau_{i,j}$ is constrained by inequality (7.20).

$$\tau_{min} \leq \tau_{i,j} \leq \tau_{max} \tag{7.20}$$

The variable $\tau_{i,j}$ is added in the same manner as described in the AS algorithm. Here τ_{min} and τ_{max} are the minimum and maximum allowed for the pheromone values respectively, these values are selected by the user but τ_{max} may be set at $1/\rho L^*$ where L^* is the minimum tour length. Clearly this is only available if the optimum tour length, or a good approximation to it, is known.

Another more recent modification of the ACO algorithm was suggested by Yu et al. (2014). They introduce chaotic behavior using the logistic map to introduce a random element in the manner in which the pheromone updates are performed. Yu et al. (2014) test their algorithm on a selection of standard test problems and conclude that their algorithm provides significant improvement on other ACS algorithms.

The ACS algorithm is naturally applied to TSP problems and a wide range of combinatorial problems such as job scheduling, knapsack problems, vehicle scheduling and many others and solutions often involve integer values. Although these problems are similar to nonlinear global optimization problems in their difficult nature, the problems tackled are different in structure. Consequently we do not provide numerical studies for the algorithm, but leave it to the reader to decide if they wish to study the algorithm in more detail using, for example, algorithms implemented by Mathworks.

7.8 SOME APPLICATIONS OF ANT COLONY OPTIMIZATION

The major application of ACO is to the TSP. It is reported that the algorithm has allowed the solution of much larger TSP problems than before its introduction. However the method has been applied to standard non-linear optimization problems as well and this approach has been described in Jalali et al. (2005). They show how the ACO algorithm, normally associated with the TSP, may be applied to the optimization of standard non-linear both unconstrained and constrained test functions. They also considered the practical problem of optimizing reservoir operations to find an optimum pattern of water releases.

Automatic programming is the use of search techniques to find programs that solve a problem. The most commonly used technique is genetic programming, which uses genetic algorithms to carry out the search. Green et al. (2004) introduce Ant Colony Programming (ACP) which uses an ant colony based search in place of genetic algorithm to solve this problem. They tested and compared it with other approaches in the literature.

Abraham et al. (2013) use ACO to attempt to find numerical solutions of Diophantine equations, an established and difficult problem in pure mathematics since there are no general methods to find solutions of such equations. Their experimental results compare well with those of other machine intelligence techniques, thereby validating the effectiveness of their proposed method.

7.9 SUMMARY

We have considered two algorithms in this chapter; the Artificial Bee Colony (ABC) algorithm and Ant colony Optimization (ACO). These algorithms share a common concept of swarm intelligence, that is by allowing the transfer and use of information between all members of the colony. In the ACO case this is achieved by using pheromones and in the ABC algorithm by using onlooker bees. The Ant Colony optimization algorithm has a long established role amongst nature inspired optimization methods having been introduced in 1991 and subsequently used to successfully solved the TSP problem for large number of cities and over the years has been applied to many demanding and practical combinatorial problems with many reported successes. The algorithm success has provided a spur to further research in the area. We have described the algorithm and provided a description of some of the developments in the field.

The ABC algorithm is a relatively recent entry to the field of nature inspired optimization the algorithms having been introduced in 2005. Specifically designed for the global optimization of non-linear optimization problems, it has been extensively studied and tested on a wind range standard test problems and practical applications. It has been reported that its performance is strong in relation to many algorithms in the field. We have described the basic algorithm and some of the suggested improvements to the basic algorithm.

7.10 PROBLEMS

7.1 For Table 7.4 in this chapter, which provides the results for Rastrigins function for various dimensions using the ABC algorithm of that problem, draw a graph of the accuracy of each average result against the dimension of the problem. Compute the error of the solution of the problem by noting the optimum solution of the problem is zero.

7.2 Use (7.11) given by Gao and Liu to generate chaotically an initial set of values for Nectar/Food locations for the ABC algorithm. In (7.11) take $K = 4$ and plot each generation of points for $k = 1, 2, 3$. Take only two points so $j = 1, 2$ and generate initial values for each of these points. Take $\mu = 0.5$, $c_{0,1} = 0.10$, $c_{0,2} = 0.25$ and $c_{0,3} = 0.05$.

7.3 Using (7.2) and (7.3) compute the onlooker bee selection probabilities, for four bees (numbered 1, 2, ... 4) located at [2, 5], [3, 2], [1, 1], [6, 2] and assume an objective function $f(x) = x_1^2 + \cos(x_2)$. Find the sum of the probabilities.

7.4 By defining $x_{ij} = 1$ as the decision if an edge i, j is traversed or $x_{ij} = 0$ if it is not traversed and L_{ij} as the length of the edge joining the towns i and j find an expression for the minimization function. Formulate the TSP as a constrained minimization problem. You should model the constraints as imposing the requirements that each city is visited only once by the salesman and each city must be visited.

7.5 This problem uses the formula developed by Goss et al. from a study of ant colony behavior for the probability p_1 of selecting a particular branch from a choice of two:

$$p_1 = \frac{(m_1 + k)^h}{(m_1 + k)^h + (m_2 + k)^h}$$

Take $k = 20$ and $h = 2$ as suggested by Goss et al. For the purposes of this study assume that 2 ants have already selected the first branch and 4 ants have selected the second branch. Use this formula 4 times for 4 different ants to calculate the probability that an ant will take a particular branch. In each case if it is greater than 0.5 add 1 to the total m_1 of ants taking the first branch else add 1 to the m_2 total.

7.6 Given an ant $k = 2$, in city $i = 3$, use (7.17) to calculate the probability that this ant will select the route from city $i = 3$ to city $j = 5$, where the set of allowed cities for this ant is [2, 4, 5]. You may assume $\alpha = 1$ and $\beta = 2$ and that the required pheromone levels are given by [0.5, 1.5, 2] and the visibility values [0.3, 0.5, 0.05].

CHAPTER 8

Physics Inspired Optimization Algorithms

8.1 INTRODUCTION

There are several optimization methods that are loosely inspired by physical phenomena. Fister et al. (2013) list over 70 nature inspired optimization methods, of which 15 are described as being inspired by Physics and Chemistry. In this chapter we consider four optimization techniques that are inspired by physical processes. Inspired is a well chosen word in this context. In biologically inspired algorithms, some degree of intelligence is present in the biological process that seeks an optimum solution. For example, in the process of ants and bees locating food. However, physics inspired processes are just based on specific laws of physics. When a ball is struck by a racket, the ball flies away from the racket, not because the ball is using its intelligence to escape from the racket but because the ball is behaving according to the laws of dynamics. In spite of this lack of innate intelligence, the methods we now describe do provide useful optimization algorithms.

In material science, annealing is a heat treatment process which involves heating material to above its recrystallization temperature, holding it at a suitable temperature and cooling it slowly to a normal temperature. In annealing, atoms migrate in the crystal lattice and the number of dislocations decreases, leading to an increase in ductility and a reduction in hardness. In terms of optimization, dislocations in the crystal lattice have been reduced. The Simulated Annealing optimization algorithm is inspired by this process.

The big bang-big crunch (BB-BC) algorithm is very loosely inspired by the expansion and the subsequent collapse of the universe. Whilst the big bang and the subsequent expansion of the universe is widely accepted as a fact, the idea that the universe will ultimately stop expanding and then collapse to a singularity is not an accepted fact. Currently the universe is still expanding, and at an accelerating rate. However, none of this affects the BB-BC algorithm, since it doesn't rely on any astrophysics process.

The gravitational search algorithm (the GSA) is very, very loosely based on the law that relates the gravitational force of attraction between two or more bodies. Indeed the method has been criticized by Gauci et al. (2012) on the grounds that it does not implement the laws of gravity correctly. All this may be true, but the fact remains that whatever the algorithm is called, it does find optimal solutions to many problems very efficiently.

Finally, Central Force Optimization (CFO), which is also based on the gravitational force between masses, but CFO uses the laws of gravitational attraction and motion in

Introduction to Nature-Inspired Optimization
DOI: 10.1016/B978-0-12-803636-5.00008-6

a somewhat different manner from the gravitational search algorithm. A striking feature of the CFO algorithm is that it is deterministic.

We now describe each of these algorithms in detail.

8.2 SIMULATED ANNEALING

Simulated annealing optimization (SA) is inspired by the process of annealing metal. Annealing involves heating and the subsequent cooling of a metal to alter its physical properties by causing changes in its internal structure. If the metal is allowed to cool sufficiently slowly, its atoms automatically align themselves into a minimum energy state, causing the ductility and hardness of the metal to change. If, however, the metal is cooled quickly, say by quenching in water, then this minimum energy state is not found. This concept of a natural process finding a minimum energy state can be utilized to find the global optima of a given non-linear function. SA was initially developed by Kirkpatrick et al. (1983) and Černeý (1985) and is thus one of the oldest heuristic optimization algorithms.

In SA optimization a temperature variable is used to simulate the cooling process. We initially set it high and then allow it to slowly 'cool' incrementally as the algorithm runs. The analogy is not perfect but the fast cooling process may be viewed as equivalent to finding a local minimum of a given non-linear function, while the slow cooling corresponds to finding the ideal energy state or a global minimum of the function, since energy level is equivalent to cost. This slow cooling process may be implemented using the Boltzman probability distribution of energy states which plays a prominent part in thermodynamics and which has the form

$$P(E) = \exp(-E/kT) \tag{8.1}$$

where $P(E)$ is the probability of E, a particular energy state, k is Boltzman's constant and T is the temperature. This function is used to reflect the cooling process where a change in the energy level, which may be initially unfavorable, may ultimately lead to a final minimum global energy state.

This corresponds to the notion of moving out of the region of a local minimum of a non-linear function in the search for a global solution for the problem. This may result in a temporary increase in the value of the objective function, i.e., climbing out of the valley of a local minimum, although convergence to the global optimum may still occur if the adjustment to the temperature is slow enough. These ideas lead to an optimization algorithm used by Kirkpatrick et al. (1983) which has the following general structure.

Let $f(\mathbf{x})$ be the non-linear function to be minimized, where $\mathbf{x}$ is an n_{var} component vector. Then

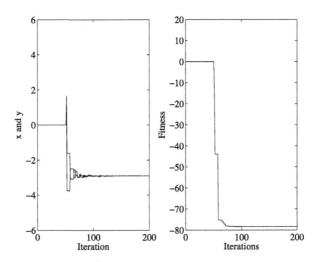

Figure 8.1 Left graph showing the changes the *x* and *y* and the right graph shows the minimum of the objective function *f* for the Styblinski-Tang function in two variables.

Step 1   Set the iteration count k and the temperature count p, to zero. Choose an initial, trial solution $\mathbf{x}^{(k)}$ and an initial, arbitrary temperature, T_p.

Step 2   Choose a second trial solution $\mathbf{x}_n$ close to $\mathbf{x}^{(k)}$

Step 3   Compute $\Delta f = f(\mathbf{x}_n) - f(\mathbf{x}^{(k)})$ then

If $\Delta f < 0$, then $\mathbf{x}_n$ replaces $\mathbf{x}^{(k)}$.

If $\Delta f > 0$ and $\exp(-\Delta f/T_p) < r$, where r is a random number in the range 0 to 1, then $\mathbf{x}_n$ replaces $\mathbf{x}^{(k)}$.

Otherwise $\mathbf{x}^{(k)}$ is unchanged.

Step 4   Set $k = k+1$. Repeat from Step 2 until there is no significant change of function value.

Step 5   Lower the current temperature using an appropriate reduction process $T_{p+1} = g(T_p)$, set $p = p+1$ and repeat from Step 2 until there is no further significant change in the function value from temperature reduction.

The key difficulties with this algorithm are choosing an initial temperature and a temperature reduction regime. This has generated many research papers but the details are not discussed here.

We illustrate graphically the use of the SA algorithm in the optimization of the Styblinski-Tang function. Figure 8.1 and Figure 8.2 show this process. The right hand plot of Figure 8.1 shows how the function value decreases, with an increasing number of iterations, to reach the minimum value. The left hand graph shows how the values of *x* and *y* change until the optimum point is reached. Note that the solution $\mathbf{x}$ does

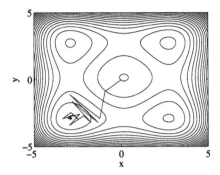

Figure 8.2 Contour plot of the Styblinski-Tang function, showing the path taken by the SA algorithm to the minimum value.

not proceed to the solution in a simple or direct manner, reflecting the random search process. This search path is shown on a contour plot of the Styblinski-Tang function, Figure 8.2.

8.3 SOME APPLICATIONS OF SIMULATED ANNEALING

Because Simulated Annealing is one of the oldest physics inspired optimization methods and it has been applied to optimize problems in many fields of science and engineering. For example, retaining walls are widely used in civil engineering and they must resist a combination of earth and hydrostatic loading. Caranic et al. (2001) describe how a modified SA algorithm was successfully applied to a constrained optimization problem to determine the minimum cost design of reinforced concrete retaining walls.

Another application is the tube hydroforming process where important parameters are the internal pressure and axial force loading paths. Theoretical calculations and trial-and-error simulations to find the optimum loading paths are both time-consuming and costly. In a paper by Mirzaali et al. (2012), pressure and force loading paths in tube hydroforming process are optimized using Simulated Annealing. The final aim is to obtain the optimal loading paths for tube hydroforming under a failure criterion. The Simulated Annealing algorithm has actually been directly incorporated into the required non-linear structural analysis.

In medical diagnostics, the analysis of data from X-ray diffraction and nuclear magnetic resonance (NMR) instrumentation generally requires sophisticated computational procedures. This leads to the non-linear optimization of a target function. In a paper by Brünger et al. (1997) simulated annealing is shown to be effective in solving this optimization problem.

8.4 THE BIG BANG-BIG CRUNCH ALGORITHM

This optimization algorithm was developed by Erol and Eksin (2006). As its name suggests the method was inspired by one of the theories of the evolution of the universe, namely that about 13.8 million years ago there was the so called "big bang" and the universe came into existence. Since then the universe has expanded, but some theoreticians think that this expansion will ultimately cease and the universe will contract until it collapses to a single point, the "big crunch". The big bang and subsequent expansion and contraction may then be repeated.

The algorithm has proved to be attractive because it is simple to implement in software and gives fast convergence and hence a low computation time. The algorithm as described seeks a global minimum of the objective function. It begins with a big bang in which the initial candidate solutions, called agents, are randomly distributed over the search space. This random placement of agents across the search space is followed by the big crunch in which the agents are drawn together to a single, representative point by computing a weighted average of the fitness of the agents. For the purposes of the description of the algorithm, we will call the process of a big bang followed by a big crunch to be an epoch. This first epoch is followed by subsequent epochs in which new sets of agents are created, taken from a *normal* distribution of random numbers around the weighted mean of the previous epoch with a standard deviation that decreases with each passing epoch. Epoch follows epoch further until convergence is achieved. We now examine the stages of the algorithm in detail.

The initial big bang randomly populates the search space with agents. Let $\mathbf{x}$ be a vector of length n_{var} elements where n_{var} is the number of variables in the problem. Let $\mathbf{x}_{max}$ and $\mathbf{x}_{min}$ be the vectors of maximum and minimum values respectively of each variable, and let $\mathbf{x}_j$ be the vector describing the position of the jth agent or candidate solution. Then the elements of $\mathbf{x}_j$ may be generated from

$$\mathbf{x}_j = \mathbf{x}_{min} + \mathbf{r}_u \circ (\mathbf{x}_{max} - \mathbf{x}_{min}), \; j = 1, \, 2, \, ...n_{pop} \tag{8.2}$$

Here n_{pop} is the number of agents, $\mathbf{r}_u$ is a vector of uniformly distributed random numbers in the range 0 to 1 and $\circ$ indicates the Hadamard or Schur multiplication. For example if $\mathbf{a}$ and $\mathbf{b}$ are vectors of three elements then

$$\mathbf{a} \circ \mathbf{b} = \begin{bmatrix} a_1 \\ a_2 \end{bmatrix} \circ \begin{bmatrix} b_1 \\ b_2 \end{bmatrix} = \begin{bmatrix} a_1 b_1 \\ a_2 b_2 \end{bmatrix} \tag{8.3}$$

From the above equation we see that each $\mathbf{x}_i$ is randomly chosen but lies in the range of the search space in each dimension.

Replace the search population by a single representative value—the big crunch. In this phase the weighted average of the candidate solutions is calculated, and denoted by the vector $\mathbf{x}_{cen}$, where

$$x_{i(cen)} = \frac{\sum_{j=1}^{n_{pop}} x_{ij}/f_j}{\sum_{j=1}^{n_{pop}} 1/f_j} \quad i = 1, 2, \dots n_{var} \tag{8.4}$$

where f_j is the value of the function at $\mathbf{x}_j$. Note that because we are seeking a minimum value of the function the algorithm uses the reciprocal of the function as a measure of fitness. Thus the weighted mean of the trial candidates is biased towards a region where the function has a minimum value. Notice also that the only information passed between epochs is $\mathbf{x}_{cen}$, a vector of only n_{var} elements.

If the value of the function is zero at some chosen location, then (8.4) cannot be evaluated due to a division by zero. Examining (8.4) under these conditions, we note that if the function value f_k tends to zero then $1/f_k$ and x_{ik}/f_k will both tend to infinity and will be the dominant terms in the summations in both the denominator and numerator of (8.4) respectively. Thus, under these circumstances (8.4) becomes

$$x_{i(cen)} = \frac{x_{ik}/f_k}{1/f_k} = x_{ik} \quad i = 1, 2, \dots n_{var}$$

If two or more values of the function are zero, then these terms will dominate the summations in (8.4). For example, if both f_k and f_l tend to zero, then

$$x_{i(cen)} = \frac{x_{ik}/f_k + x_{il}/f_l}{1/f_k + 1/f_l}$$

Now f_k and f_l tend to zero identically and so

$$x_{i(cen)} = \frac{(x_{ik} + x_{il})/f_k}{(1+1)/f_k} = (x_{ik} + x_{il})/2 \quad i = 1, 2, \dots n_{var}$$

In general $x_{i(cen)}$ will be the mean value of the locations where the function value is zero.

The following big bang expansion provides agents for the next epoch. Then new agents are generated so that they are normally distributed around the weighted average of the previous epoch agents or candidate solutions. Note that in this algorithm, the agents of one epoch are not directly related to the agents of the previous epoch, indeed, we could choose to have a different number of agents for each epoch, although this would serve no useful purpose and add to the complexity of programming. Thus the big bang expansion takes the form

$$x_{ij}^{(k+1)} = x_{i(cen)}^{(k)} + \left(\frac{\alpha}{k}\right) r_n \left(x_{i(max)}^{(k)} - x_{i(min)}^{(k)}\right) \tag{8.5}$$

where $i = 1, 2, ..., n_{var}, j = 1, 2, ..., n_{pop}$ and r_n is a random number generated from a normal distribution of numbers with a mean of zero and a standard deviation equal to one. By multiplying the random number by α/k we change the standard deviation of the set of random numbers to α/k. Thus α is a parameter to limit the size of the search space and it is divided by the epoch number, k, thereby reducing the size of the search space with the increasing number of epochs.

The above equation can be modified to introduce the influence of the current global best solution on the location of the new agent as suggested by Yesil and Urbas (2010) as follows

$$x_{ij}^{(k+1)} = \beta x_{i(cen)}^{(k)} + (1 - \beta) x_{i(best)}^{(k)} + \left(\frac{\alpha}{k}\right) r_n \left(x_{i(max)}^{(k)} - x_{i(min)}^{(k)}\right) \quad i = 1, 2, ..., n_{var} \tag{8.6}$$

where $x_{i(best)}^{(k)}$ is the best of the trial candidates or agents of the previous epoch and β is a parameter controlling the influence of the best global solution from the previous epoch. This equation may the written

$$\mathbf{x}_j^{(k+1)} = \beta \mathbf{x}_{cen}^{(k)} + (1 - \beta) \mathbf{x}_{best}^{(k)} + \left(\frac{\alpha}{k}\right) \mathbf{r}_n \left(\mathbf{x}_{max}^{(k)} - \mathbf{x}_{min}^{(k)}\right) \tag{8.7}$$

where $\mathbf{x}_{best}^{(k)}$ is the best of the trial candidates or agents of the previous epoch.

Recombination. We do not take $\mathbf{x}_j^{(k+1)}$ as our new agents but we compare $\mathbf{x}_j^{(k+1)}$ with $\mathbf{x}_j^{(k)}$ and choose the one which gives the lower function value, i.e. the better one since this is a minimization procedure, as follows:

$$\mathbf{x}_{j(next)} = \begin{cases} \mathbf{x}_j^{(k)} & \text{if } f(\mathbf{x}_j^{(k)}) \leq f(\mathbf{x}_j^{(k+1)}) \\ \mathbf{x}_j^{(k+1)} & \text{otherwise} \end{cases} \tag{8.8}$$

This concludes the description of the standard BB-BC algorithm.

Some modifications and improvements have been suggested to the standard BB-BC algorithm. One weakness of the BB-BC algorithm is that it can be trapped in a local optimum. To avoid this Kaveh and Talatahari (2009) developed a hybrid version of the BB-BC. This was done by combining the BB-BC algorithm with features of Particle Swarm Optimization (PSO) (Kennedy and Eberhart, 1995).

Further modifications to the basic BB-BC algorithm were introduced by Hasançebi and Azad (2013) in a paper concerning the optimization of structures. This is a discrete

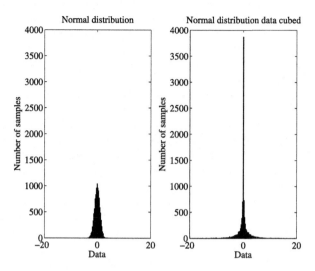

Figure 8.3 Histogram of data sampled from a normal distribution, and data sampled from a normal distribution and then cubed.

optimization problem and (8.5) is replaced by

$$d_{ij}^{(k+1)} = d_{i(cen)}^{(k)} + \text{round}\left[\left(\frac{\alpha}{k}\right) r_n \left(d_{i(max)}^{(k)} - d_{i(min)}^{(k)}\right)\right] \tag{8.9}$$

where d_{ij} is a discrete, integer design variable, in contrast to x_{ij} of (8.5) which is a real or continuous variable. The round or rounding function is introduced to ensure that the variable $d_{ij}^{(k+1)}$ remains an integer value. Hasançebi and Azad then consider two alternative random distributions for the above equation. The first, which they describe as the modified BB–BC algorithm replaces (8.9) by

$$d_{ij}^{(k+1)} = d_{i(cen)}^{(k)} + \text{round}\left[\left(\frac{\alpha}{k}\right) r_n^p \left(d_{i(max)}^{(k)} - d_{i(min)}^{(k)}\right)\right] \tag{8.10}$$

In this equation, the random number, taken from a normal distribution is raised to the power p, where $p > 2$ and must be an odd number. The effect of raising r_n to the power 3 is shown in Figure 8.3.

When r_n is less than 1, r_n^p is smaller than r_n and data are clustered closer to zero. In contrast, when r_n is greater than 1, r_n^p is larger than r_n and data are spread to higher values. This means that some trial values based on this distribution are close to the mean value computed from the previous epoch and help exploitation, where as other points are widely spread and help exploration to avoid the process getting trapped in local optima.

Table 8.1 20 runs of BB-BC: ROS2. $n_{gen} = 500$, $n_{pop} = 20$, $\alpha = 0.5$, $\beta = 0.1$, $x = -5$ to 5

	Mean	Best	Worst	St Dev
Parameters above	8.2294×10^{-7}	7.0567×10^{-9}	2.5889×10^{-6}	6.8550×10^{-7}
$n_{pop} = 40$	2.4180×10^{-7}	1.2486×10^{-9}	9.9881×10^{-7}	2.4277×10^{-7}
$n_{gen} = 200$	1.2265×10^{-5}	1.8009×10^{-7}	4.7978×10^{-5}	1.0877×10^{-5}
$n_{gen} = 1000$	5.9909×10^{-8}	6.2007×10^{-9}	2.3483×10^{-7}	5.4210×10^{-8}
$\alpha = 0.3$	2.1305×10^{-7}	1.5213×10^{-10}	1.1936×10^{-6}	2.7967×10^{-7}
$\alpha = 0.7$	1.4241×10^{-6}	6.1414×10^{-8}	4.1848×10^{-6}	1.1717×10^{-6}
$\beta = 0.05$	7.3063×10^{-7}	2.1351×10^{-8}	1.8938×10^{-6}	5.7516×10^{-7}
$\beta = 0.2$	4.2146×10^{-7}	1.6021×10^{-8}	1.4628×10^{-6}	3.7383×10^{-7}

The second modification is to replace the normal distribution by the exponential distribution, again raised to the power p. The exponential distribution probability density function is

$$E(x) = \lambda e^{\lambda x}$$

where $x > 0$.

$$d_{ij}^{(k+1)} = d_{i(cen)}^{(k)} + \text{round}\left[\left(\frac{\alpha}{k}\right) r_e^p \left(d_{i(max)}^{(k)} - d_{i(min)}^{(k)}\right)\right] \tag{8.11}$$

with λ in the exponential distribution equal to 1 and p again equal to 3. Unlike a normal distribution, the exponential distribution provides only positive real numbers. Therefore in (8.11) the rounded terms must be added or subtracted from $d_{i(cen)}^{(k)}$ with equal probability to allow for both an increase and decrease in $d_{ij}^{(k+1)}$.

The basic BB–BC algorithm may be summarized thus:

BOX 8.2 Summary of the BB-BC algorithm

Step 1: Generate an initial population of agents, located in the search space from a uniform distribution of random numbers, (8.2).

Step 2: Determine fitness of the function at each of the agent locations

Step 3: Determine the location of the mean value, weighted by fitness, of the agents using (8.4).

Step 4: Generate a new population of agents, using a normal distribution of random numbers centered at the weighted mean value of the previous generation of agents using (8.5), and reducing the standard deviation of these random numbers at each successive epoch.

Step 5: If the required number of epochs is complete then end, else go to Step 2.

8.5 SELECTED NUMERICAL STUDIES USING THE BB-BC ALGORITHM

Table 8.1 shows the optimization of Rosenbrock's function in 2 dimensions in the search range x, $y = -5$ to 5. The first row of the table uses the values of the parameter

Table 8.2 20 runs of BB-BC: S-T2. $n_{gen} = 200$, $n_{pop} = 20$, $\alpha = 0.5$, $\beta = 0.1$, $x = -5$ to 5

	Mean	Best	Worst	St Dev
Parameters above	−75.5049762	−78.3323314	−64.1955768	5.80158849
$n_{pop} = 40$	−76.9186502	−78.3323309	−64.1955732	4.35119411
$n_{gen} = 100$	−76.9185551	−78.3323308	−64.1954001	4.35120025
$n_{gen} = 500$	−75.5049868	−78.3323313	−64.1956105	5.80158728
$\alpha = 0.3$	−74.0913105	−78.3323313	−64.1956010	6.64655378
$\alpha = 0.7$	−77.6254713	−78.3323313	−64.1955790	3.16106864
$\beta = 0.05$	−76.2118056	−78.3323303	−64.1955961	5.17894938
$\beta = 0.2$	−74.7981391	−78.3323305	−64.1955940	6.28040088

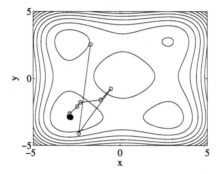

Figure 8.4 Contour plot of the Styblinski-Tang function, showing the path taken by the weighted mean of the BB-BC algorithm to the minimum value, starting near to the point $(3, -2)$.

indicated in the title. The table shows that increasing the population of agents from 20 to 40 marginally improves the results, and increasing the number of generations improves the results significantly. Conversely decreasing the number of generations significantly degrades the results. In this problem decreasing α has provided an improved result. In contrast changing the value of β has little effect.

Table 8.2 shows an analysis of the optimization of Styblinski–Tang function in 2 dimensions. This problem has four minima and in this case changes to the parameters α and β have some effect on the mean but little effect on the best results. Increasing the population of agents from 20 to 40 gives some improvement in the result.

Figure 8.4 shows the path taken by $\mathbf{x}_{cen}$, the weighted mean of the population. It is seen that it converges to $x, y = -2.9035$, the location of the minimum value of the function. Figure 8.5 shows the path of $\mathbf{x}_{cen}$ as the algorithm searches for the minimum in the region of $x, y = -2.9035$.

A more demanding test of the BB-BC algorithm is to find the minimum of Rastrigin's in 6 variables which has a minimum of zero. The results are shown in Table 8.3. For this particular problem with the chosen parameters, increasing the number of gen-

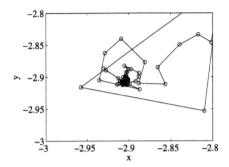

Figure 8.5 A close up of the region close to the minimum value of the Styblinski-Tang function, showing the path taken by the weighted mean of the BB-BC algorithm.

Table 8.3 20 runs of BB-BC: RAS6. $n_{gen} = 500$, $n_{pop} = 20$, $\alpha = 0.5$, $\beta = 0.1$, $x = -5$ to 5

	Mean	Best	Worst	St Dev
Parameters above	11.8965684	2.00087135	23.8932414	5.80677082
$n_{pop} = 40$	15.7267636	3.98297833	33.8348403	7.89038441
$n_{gen} = 1000$	11.7417724	2.98627194	34.8238962	7.85140884
$n_{gen} = 2000$	11.8402491	1.99010643	26.8641371	6.22521282
$n_{gen} = 4000$	12.5862501	3.97986490	21.8891010	5.55633987
$\alpha = 0.3$	13.5338601	3.98204502	33.8307499	7.51805221
$\alpha = 0.7$	11.8515543	3.99331804	21.9029955	5.60974853
$\beta = 0.05$	12.1443827	1.00414640	27.8657126	8.65870141
$\beta = 0.2$	11.4982946	1.00060307	25.8746816	6.63528944

Table 8.4 20 runs of BB-BC: RAS. $n_{gen} = 2000$, $n_{pop} = 40$, $\alpha = 0.5$, $\beta = 0.1$, $x = -5$ to 5

n_{var}	Mean	Best	Worst	St Dev
2	0.4477317	1.18×10^{-8}	1.9899183	0.6017566
4	3.5818800	0.99498323	6.9647667	2.0003618
6	11.0442169	2.98512806	22.8841616	5.6918379
8	14.6761693	6.96540120	25.8692970	5.3029847
10	24.3278504	2.98628889	44.7741643	10.1899839
12	28.6567501	6.96651172	46.7651500	12.3369509

erations does not improve the result. Both increasing and decreasing the value of β has given a better result in this particular case, although this may just be due to the random nature of the algorithm. Table 8.4 shows the accuracy falls off as the number of variables is increased, as might be expected.

8.6 SOME APPLICATIONS OF BIG BANG-BIG CRUNCH OPTIMIZATION

Tabrizian et al. (2013) explore a damage assessment methodology based on changes in dynamic properties of a structural system. A procedure for locating and quantifying damaged areas in the structure based on Big Bang-Big Crunch (BB-BC) optimization was developed. To verify the method a number of damage scenarios for simulated structures were considered. BB-BC optimization in damage detection was compared with particle swarm optimization. Tabrizian et al. show that the BB-BC optimization method is a feasible methodology to detect damage location and its severity while introducing numerous advantages compared to other methods.

Uppal and Kumar (2016) have used the Big Bang-Big Crunch (BB-BC) to study the deployment of wireless sensor networks, a factor that significantly affects their performance. Uppal et al. claim that compared with the ABC approach, the BB-BC the approach is much better.

8.7 THE GRAVITATIONAL SEARCH ALGORITHM

Gravitational Search Algorithm (GSA) was first proposed by Rashedi et al. (2009). It is a heuristic approach for determining the optimum solution of a function and the method is very broadly based on a simulation of Newton's law of gravity. This law states that every particle in the universe attracts every other particle with a force that is directly proportional to the product of their masses and inversely proportional to the square of the distance between them. Stated mathematically we have

$$F_{12} = G\frac{m_1 m_2}{d_{12}^2} \qquad (8.12)$$

where m_1 and m_2 are the masses of two particles a distance d_{12} apart, G is the gravitational constant and F_{12} is the force of attraction acting on each particle, attracting it towards the other particle. The acceleration, a_1 of the mass m_1, acted on by a force F_{12} is $a_1 = F_{12}/m_1$ and so the acceleration of mass m_1 towards mass m_2, by virtue of (8.12), is

$$a_1 = G\frac{m_2}{d_{12}^2} \qquad (8.13)$$

We can consider the GSA to be based an isolated system of particles or agents attracted to each other by a force which is similar, but not identical, to a gravitational force. The mass of each agent is related to the fitness of the function we seek to optimize. Consider a system of n_{pop} masses, m_i, where $i = 1, 2, ... n_{pop}$, and let the position of mass m_i be $\mathbf{x}_i$.

Assume we wish to determine the minimum value of a function. At time t we define $m_i^{\prime(t)}$, located at $\mathbf{x}_i^{(t)}$ by

$$m_i^{\prime(t)} = \frac{f_i^{(t)} - f_{max}^{(t)}}{f_{min}^{(t)} - f_{max}^{(t)}} \tag{8.14}$$

In this equation $f_i^{(t)}$ is the function value at location $\mathbf{x}_i^{(t)}$. We see that if $f_i^{(t)} = f_{max}^{(t)}$ then $m_i^{\prime(t)} = 0$ and if $f_i^{(t)} = f_{min}^{(t)}$ then $m_i^{\prime(t)} = 1$. Thus the maximum value of $m_i^{\prime(t)}$ is associated with the minimum value of $f_i^{(t)}$. Note that, alternatively, if we wish to determine the maximum of the function then $m_i^{\prime(t)}$ is defined by

$$m_i^{\prime(t)} = \frac{f_i^{(t)} - f_{min}^{(t)}}{f_{max}^{(t)} - f_{min}^{(t)}}$$

where $f_{min}^{(t)}$ is the minimum of $f_i^{(t)}$, $i = 1, 2 ... n_{pop}$ and $f_{max}^{(t)}$ is the maximum of $f_i^{(t)}$. We relate the mass of the ith particle $m_i^{(t)}$ to the function fitness using

$$m_i^{(t)} = \frac{m_i^{\prime(t)}}{\sum_{j=1}^{n_{pop}} m_j^{\prime(t)}} \tag{8.15}$$

Thus $m_i^{(t)}$ is proportional to $m_i^{\prime(t)}$. Having determined the particle masses at the location at time t then Rashedi et al. (2009) suggest that the force acting between particles i and k at time t is given by

$$\mathbf{F}_{ik}^{(t)} = G^{(t)} \frac{m_i^{(t)} m_k^{(t)}}{d_{ik} + \epsilon} (\mathbf{x}_k^{(t)} - \mathbf{x}_i^{(t)}) \tag{8.16}$$

where $d_{ik} = ||\mathbf{x}_i, \mathbf{x}_k||_2$, the Euclidean distance between the two agents or masses. Note that this is equivalent to stating that $F = G m_1 m_2$, i.e. the gravitational force is independent of the distance between the particles and depends only on the product of their masses and $G^{(t)}$ is not constant but varies with time. Typically, $G^{(t)}$ is defined by

$$G^{(t)} = G_0 e^{(-\alpha t / t_{max})} \tag{8.17}$$

We now use Newton's law of motion relating an applied force and the resultant acceleration of a mass. Thus the acceleration of mass i due to the influence of mass k is

$$\mathbf{a}_{ik}^{(t)} = \mathbf{F}_{ik}^{(t)} / m_i^{(t)} = G^{(t)} \frac{m_k^{(t)}}{d_{ik} + \epsilon} (\mathbf{x}_k^{(t)} - \mathbf{x}_i^{(t)}) \tag{8.18}$$

We introduce a random element in to the simulation, by supposing that the acceleration of m_i due to all the other agents is

$$\mathbf{a}_i^{(t)} = \sum_{k=1, k \neq i}^{n_{pop}} \mathbf{r}_u \circ \mathbf{a}_{ik}^{(t)} \tag{8.19}$$

where $\mathbf{r}_u$ is a vector of random numbers with a uniform distribution in the range 0 to 1.

Having determined the acceleration of each mass or agent we can derive its velocity and change in position. A further random element is introduced in the velocity of the particles at time $t + 1$ as follows

$$\mathbf{v}_i^{(t+1)} = \mathbf{r}_u \circ \mathbf{v}_i^{(t)} + \mathbf{a}_i^{(t)} \tag{8.20}$$

where $\mathbf{r}_u$ is a vector of uniformly distributed random numbers in the range 0 to 1. Thus, the new position of the mass or agent, at time $t + 1$ becomes

$$\mathbf{x}_i^{(t+1)} = \mathbf{x}_i^{(t)} + \mathbf{v}_i^{(t+1)} \tag{8.21}$$

Having determined the location of all the agents or masses at time $t + 1$ we can use (8.15) and (8.16) to determine a new set of forces and repeat the process.

BOX 8.3 Summary of the GSA

Step 1: Generate an initial population of agents, located in the search space from a uniform distribution of random numbers.

Step 2: Determine fitness of the function at each of agent location and determine the equivalent mass from (8.14) and (8.15).

Step 3: Determine the forces of attraction between each agent using (8.16) where the gravitational factor reduces according to (8.17).

Step 4: Determine the acceleration of each agent due to another agent, see (8.18).

Step 5: Determine the acceleration of each agent and introduce a random element using (8.19), and thus the velocity, with a randomness, (8.20), and hence the new positions of the agents (8.21).

Step 6: If the required number of iterations is complete then end, else update count and go to Step 2.

This concludes the description of the GAS algorithm. Some modifications to the basic algorithm have been proposed. One of which is to combine the GSA with PSO (Gu and Pan, 2013).

8.8 SELECTED NUMERICAL STUDIES USING THE GSA

Table 8.5 shows the results of applying the GSA to the Styblinski-Tang function. Increasing both the number of generations and population size improves the result. The Table shows that the mean, worst and best minimum values are accurate to at least 4 decimal places when the population is 20 using 1000 generations.

Table 8.6 shows the solution of Rastrigin's function in six variables. Clearly, increasing the number of generations provides some improvement to the accuracy of the

Table 8.5   20 runs of GSA: S-T2. $G_0 = 100$, $\alpha = 20$, $\epsilon = 10^{-4}$, $x = -5$ to 5

pop	gens	Mean	Best	Worst	St Dev
20	500	−75.5049876	−78.3323314	−64.1956121	5.80158711
40	500	−77.6254954	−78.3323314	−64.1956124	3.16106647
20	200	−72.6776438	−78.3323314	−64.1956124	7.10546403
40	200	−76.2118235	−78.3323314	−64.1956124	5.17895237
20	1000	−78.3323314	−78.3323314	−78.3323312	4.10×10^{-8}

Table 8.6   20 runs of GSA: RAS6. $n_{gen} = 500$, $G_0 = 2$, $\alpha = 20$, $n_{pop} = 20$, $\epsilon = 10^{-4}$, $x = -5$ to 5

	Mean	Best	Worst	St Dev
Parameters above	3.08437233	0.99495906	5.96974931	1.70508175
$n_{gen} = 1000$	3.13412055	0.99495906	5.96974932	1.37997689
$n_{gen} = 2000$	2.48739718	0.99495906	7.95966742	1.75329128
$G_0 = 1$	3.08437259	0.99495906	5.96974935	1.67424615
$G_0 = 20$	3.88033882	0.99495906	8.95462648	2.03905741
$\alpha = 10$	2.58689305	5.55×10^{-11}	5.96974931	1.49330991
$\alpha = 40$	2.03966802	1.18×10^{-9}	3.97983624	1.18496679
$n_{pop} = 40$	1.44269066	0.99495906	2.98487717	0.75532778

Table 8.7   20 runs of GSA: RAS6. $G_0 = 100$, $\alpha = 20$, $\epsilon = 10^{-4}$, $x = -5$ to 5

pop	gens	Mean	Best	Worst	St Dev
20	500	6.17504456	0.994959057	25.9949591	5.37370430
40	500	2.03966607	0.994959057	3.9798362	0.93975201
20	1000	4.73235599	9.990×10^{-8}	28.9798362	6.03877671
40	1000	1.94017021	0.994959057	3.9798362	0.88256968

minimum value. Increasing the size of the population also provides an improved result. However, changing the value of the parameter G_0 does not seem to have a significant effect. The parameter α, which controls the decay of the gravitational coefficient provides an improved result when increased to 40.

Table 8.7 optimizes Rastrigin's function with $G_0 = 100$. Clearly increasing the population size and the maximum number of generations gives a significant improvement in the result, but the change in the value of G_0 doesn't have much effect.

Table 8.8 shows the optimization of Rosenbrock's function in 2 variables. Increasing the number of generations provides a significant improvement in the accuracy of the result. Changing the parameters α and G_0 does not produce a significant change in the accuracy of the minimization in this example. Figure 8.6 and Figure 8.7 show the paths of 6 agents for 500 generations as they seek the minimum of Rosenbrock's function, with $G_0 = 2$ and $\alpha = 20$. A small value of G_0 has been chosen so that the steps between

Table 8.8 20 runs of GSA: ROS2. $n_{gen} = 500$, $G_0 = 2$, $\alpha = 20$, $n_{pop} = 40$, $\epsilon = 10^{-4}$, $x = -5$ to 5

	Mean	Best	Worst	St Dev
Parameters above	0.2057747	0.0059106	0.3314851	0.0635076
$n_{pop} = 20$	0.2324104	0.0345554	1.3411559	0.2636336
$n_{gen} = 200$	0.7449516	0.2325723	2.8804125	0.5733027
$n_{gen} = 1000$	0.0989385	0.0890175	0.1166128	0.0075266
$G_0 = 1$	0.4534843	0.0973805	1.8598576	5.4740793
$G_0 = 20$	0.2037003	0.1803625	0.2341688	0.0142034
$\alpha = 10$	0.1009482	0.0798391	0.1207355	0.0096850
$\alpha = 40$	0.3933877	0.0045608	1.1977795	0.2617141

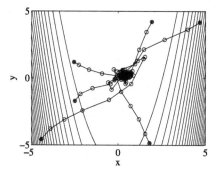

Figure 8.6 Contour plot of Rosenbrock's function showing the path of 6 randomly chosen agents, moving towards the minimum of the function at (1, 1).

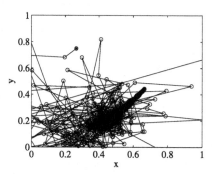

Figure 8.7 A close up of the region close to the minimum of Rosenbrock's function showing the path of 6 randomly chosen agents, moving towards the minimum of the function at (1, 1).

each iteration are small and we obtain relatively smooth paths taken by the agents as they converge. Had a larger value of G_0 been taken the paths taken by the agents would be more akin to the path of the probes of Central Force Optimization shown in

Table 8.9 20 runs of GSA: ROS2. $G_0 = 2, \alpha = 20, \epsilon = 10^{-4}, x = -5$ to 5

pop	gens	Mean	Best	Worst	St Dev
20	500	1.66612755e-01	7.13115776e-04	2.55822812e-01	5.89148449e-02
40	500	1.97493672e-01	6.93519892e-02	2.28243929e-01	3.67741135e-02
20	1000	8.13583731e-02	5.92095411e-02	9.67277408e-02	1.16401409e-02
40	1000	9.74436880e-02	7.51626953e-02	1.19191880e-01	9.84188403e-03

Table 8.10 20 runs of GSA: RAS. $n_{gen} = 2000, n_{pop} = 40, G_0 = 100, \alpha = 20,$ $\epsilon = 10^{-4}, x = -5$ to 5

n_{var}	Mean	Best	Worst	St Dev
2	0.39798364	2.9310×10^{-13}	1.98991812	0.59522733
4	1.14420292	4.6025×10^{-12}	2.98487718	1.03474735
6	1.84067443	2.0966×10^{-11}	2.98487908	0.98310613
8	3.18386896	1.3342×10^{-9}	6.96470839	1.50027234
10	3.73109651	0.994959051	6.96471340	1.36479189
12	5.52202155	1.989918111	8.95462668	1.97612271

Figure 8.13. Note also that Figure 8.7 shows that the 6 agents do not to converge to the minimum at $x = y = 1$. Thus, this is a poor result, but it is to be expected since only 6 agents are used.

Table 8.9 shows the significance of increasing the number of generations in finding the solution. However, increasing the population size does not produce an improved result.

Table 8.10 shows that for a given number of generations and population size, the performance of the algorithm deteriorates as the number of variables increases.

8.9 SOME APPLICATIONS OF THE GRAVITATIONAL SEARCH ALGORITHM

The GSA algorithm has been applied by Rashedi et al. (2011) to the design of linear and non-linear filters. For example, to the design of infinite impulse response (IIR) filters. The parameters of the filters constitute a vector to be optimized. The results they obtain demonstrate the efficiency of the GSA method.

Parameter identification of hydraulic turbine governing systems (HTGS) is crucial in modeling hydro-power plants. In a paper by Li and Zhou (2011) the gravitational search algorithm (GSA) is applied in parameter identification of the HTGS and is validated by comparing experimental and simulated results.

Clustering is used to group data objects into sets of disjoint classes called clusters so that objects within the same class are highly similar to each other and dissimilar from the objects in other classes. K-harmonic means (KHM) is one of the most popular clustering techniques, and has been applied widely, but this method may obtain the local rather than the global optimum. A hybrid data clustering algorithm, based on an improved version of GSA and KHM, is proposed by Yin et al. (2011) with the merits of both algorithms. This not only helps the KHM clustering to escape from local optima but also overcomes the slow convergence speed of the GSA. The authors claim that tests on seven data sets show results that indicate that this hybrid approach is superior to other methods.

8.10 CENTRAL FORCE OPTIMIZATION

Central Force Optimization (CFO) has some similarity to the Gravitational Search Algorithm in that it inspired by gravitational kinematics and the laws of gravity. However, in contrast to every other nature inspired optimization method described in this book, this algorithm is inherently deterministic. Furthermore, the details of the implementation are very different from those of the GSA.

The algorithm models "probes" that "fly" through the multi-dimensional decision space, analogous to masses moving under the influence of gravity. Equations are developed for the probes' positions and accelerations using the analogy of particle motion in a gravitational field. In the physical universe, objects traveling through three dimensional space become trapped in closed orbits around large masses creating high gravitational fields. This is analogous to locating the maximum value of an objective function. In the CFO algorithm, "mass" is a particular value of the objective function to be *maximized*.

Beginning with Newton's law of gravity,

$$F = G\frac{m_1 m_2}{d^2} \tag{8.22}$$

where d is the distance between masses m_1 and m_2 and G is the gravitational constant. Thus

$$\mathbf{a}_{12} = -F/m_1 = G\frac{m_2}{d^2} \times \hat{\mathbf{d}} \tag{8.23}$$

where $\mathbf{a}_{12}$ is the acceleration of mass m_1 due to mass m_2, $\hat{\mathbf{d}}$ is a unit vector pointing from m_2 to m_1 along a line joining the two masses.

The position vector of a particle subject to constant acceleration during the interval t to $t + \Delta t$, is given by the standard equation for displacement in terms of initial velocity and acceleration, thus:

$$\mathbf{x}^{(t+1)} = \mathbf{x}^{(t)} + \mathbf{v}^{(t)}\Delta t + \mathbf{a}(\Delta t)^2 \tag{8.24}$$

where $\mathbf{x}^{(t+1)}$ is the position of the mass at time $t + \Delta t$, $\mathbf{x}^{(t)}$ and $\mathbf{v}^{(t)}$ are the position and velocity vectors at time t respectively and $\mathbf{a}$ is the acceleration of the mass.

We now wish to implement these ideas as an algorithm. Letting $\mathbf{a}_{ps}^{(t-1)}$ be the acceleration of probe p due to probe s at time $t-1$, Formato (2007) proposed the formula

$$\mathbf{a}_{ps}^{(t-1)} = GU\left(m_s^{(t-1)} - m_p^{(t-1)}\right) \times \left(m_s^{(t-1)} - m_p^{(t-1)}\right)^{\alpha} \frac{\left(\mathbf{x}_s^{(t-1)} - \mathbf{x}_p^{(t-1)}\right)}{\left\|\mathbf{x}_s^{(t-1)} - \mathbf{x}_p^{(t-1)}\right\|^{\beta}} \qquad (8.25)$$

where $m_s^{(t-1)}$ and $m_p^{(t-1)}$ are the mass of probes p and s at locations $\mathbf{x}_s^{(t-1)}$ and $\mathbf{x}_p^{(t-1)}$ respectively. The masses are equal to the function values at s and p, i.e. $m_s^{(t-1)} = f(\mathbf{x}_s^{(t-1)})$. The term $\|\mathbf{x}_s^{(t-1)} - \mathbf{x}_p^{(t-1)}\|$ is the Euclidean distance. The function $U(z)$ is the unit step function defined by

$$U(z) = \begin{cases} 1 & \text{if } z \geq 0 \\ 0 & \text{otherwise} \end{cases}$$

Thus $U\left(m_s^{(t-1)} - m_p^{(t-1)}\right)$ is equal to either 0 or 1. This is because, if a negative term was allowed, the gravitational force would be a force of repulsion rather than attraction. This would not accord with physical reality.

Note also that in this algorithm, Formato chose to use the difference of the mass, i.e. the difference of the fitnesses instead of the mass (fitness) values themselves because, he argued, this would avoid excessive gravitational "pull" by other very close probes. Probes that are located nearby are likely to have similar fitness values, which could lead to an excessive force on the probe being considered. The parameters α and β and G are user-defined coefficients that must be greater than zero. Typically Formato chose $\alpha = 2$, $\beta = 2$ and $G = 2$. Summing over all the probes gives

$$\mathbf{a}_p^{(t-1)} = G\sum_{\substack{s=1 \\ s \neq p}}^{n_{pop}} U\left(m_s^{(t-1)} - m_p^{(t-1)}\right) \times \left(m_s^{(t-1)} - m_p^{(t-1)}\right)^{\alpha} \frac{\left(\mathbf{x}_s^{(t-1)} - \mathbf{x}_p^{(t-1)}\right)}{\left\|\mathbf{x}_s^{(t-1)} - \mathbf{x}_p^{(t-1)}\right\|^{\beta}} \qquad (8.26)$$

Note that when we sum the contribution of each probes we exclude probe p since it cannot interact with itself. The position vector of a particle subject to constant acceleration during the interval t to $t + \Delta t$ is given in (8.24). We can arbitrarily choose to make $\Delta t = 1$. Thus using (8.24) we have

$$\mathbf{x}_p^{(t)} = \mathbf{x}_p^{(t-1)} + \mathbf{v}_p^{(t-1)} + \mathbf{a}_p^{(t-1)}/2 \qquad (8.27)$$

where $\mathbf{v}_p^{(t-1)} = (\mathbf{x}_p^{(t-1)} - \mathbf{x}_p^{(t-2)})/\Delta t$ and $\Delta t = 1$. However, Formato's experience of using this algorithm has shown that setting $\mathbf{v}_p^{(t-1)} = 0$ gives satisfactory results.

All other optimization methods described in Chapters 2 to 7 include a random element in the algorithm and consequently it is customary to run the algorithm several times (typically 10 or 20 times) and the mean, best, worst result and the standard deviation of the results are provided. In contrast, CFO is deterministic and for a given number of probes and a given set of initial probes locations, the same result will be obtained. Furthermore, it is necessary to choose the initial probe locations; they are not chosen randomly. It is obviously sensible to make the initial number of the probes identical in each dimension. A frequently chosen arrangement in each dimension is to space the probes uniformly along, or parallel to the axis. We can run CFO with different numbers of probes and different initial locations. This is normally done systematically. If we assume there are n_d probes distributed in each dimension, then the number of probes, n_{pop} is equal to $n_d n_{var}$. Suppose we want to generate a line of probes parallel to the ith axis, where $i = 1, 2, ...n_{var}$. Then we let

$$x_i = x_{i(min)} + \gamma(x_{i(max)} - x_{i(min)}), \quad i = 1, 2, ...n_{var} \tag{8.28}$$

where γ is chosen such that $0 \le \gamma \le 1$. Thus γ adjusts the position of the line of probes parallel to the axes from the lower bound ($\gamma = 0$) to the upper bound ($\gamma = 1$) of the search area. Thus the line of probes is parallel to the ith axis and passes through the point x_i. The locations of the probes along a line parallel to the ith axis, x_{ij}, is given by

$$x_{ij} = x_{i(min)} + (j - 1)(x_{i(max)} - x_{i(min)})/(n_d - 1), \quad j = 1, 2, ...n_d \tag{8.29}$$

where x_{ij} is the location of the jth probe along the line parallel to the ith axis. Obviously we can run the CFO algorithm with various values of n_p and γ.

Figure 8.8 shows uniformly spaced lines of 8 probes in a problem of three dimensions, with $\gamma = 0$ and $\gamma = 0.5$.

As the CFO algorithm proceeds, it is possible for a probe in a new iteration to move out of the search space to a region where a solution is not sought. These errant probes must be returned to the search space. There are many ways that this might be achieved but the CFO algorithm uses a simple and deterministic method to reposition errant probes, as follows:

$$\text{if } x_{ip}^{(t)} < x_{i(min)} \text{ then } x_{ip}^{(t)} = x_{i(min)}^{(t)} + F_{rep}\left(x_{ip}^{(t-1)} - x_{i(min)}\right) \tag{8.30}$$

$$\text{if } x_{ip}^{(t)} > x_{i(max)} \text{ then } x_{ip}^{(t)} = x_{i(max)}^{(t)} - F_{rep}\left(x_{i(max)} - x_{ip}^{(t-1)}\right) \tag{8.31}$$

F_{rep} is the repositioning factor. If $F_{rep} = 0$ the errant probe is moved back to the nearest limit of the search space, if $F_{rep} = 1$ the probe is moved to its previous position in the particular dimension. The repositioning factor is often chosen to be 0.5 but in the optimization of some problems different values of the repositioning factor have been used in the range $0 \le F_{rep} \le 1$. A further variant is to allow F_{rep} to change at each iteration, typically varying from an initial value of F_{rep} in increments of ΔF_{rep} to 1.

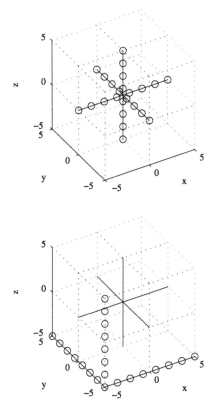

Figure 8.8 The initial distribution of probes. The upper diagram shows that with $\gamma = 0.5$ lines of 8 probes intersect at [0 0 0]. The lower diagram shows that with $\gamma = 0$ they intersect at $[-5 \ -5 \ -5]$.

Formato states that this introduces an element of pseudo-randomness into the algorithm (see Figure 8.9).

A further refinement of CFO is to reduce the search space so that it is smaller than the solution space, Formato (2013). This reduction begins (arbitrarily) at iteration step 20 and continues at step 40, 60, 80 and so on. The new search region boundary is $x'_{i(min)}$ to $x'_{i(max)}$ in each dimension i are

$$x'_{i(min)} = x_{i(min)} + 0.5 \left(x_{i(best)} - x_{i(min)}\right) \tag{8.32}$$

$$x'_{i(max)} = x_{i(max)} - 0.5 \left(x_{i(max)} - x_{i(best)}\right) \tag{8.33}$$

where $x_{i(best)}$ is the best value of x_i up to the current iteration. This reduction in the search area is illustrated in Figure 8.10 where the reduced search areas that arise in the optimization of the Styblinski-Tang function are shown. By reducing the search space convergence is likely to be faster. However, this may be at the expense of exploration.

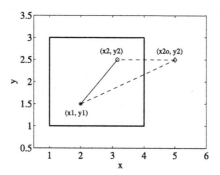

Figure 8.9 Probe initially in the rectangular search space at (x_1, y_1). It tries to move to (x_{2o}, y_2) outside the search space. To prevent this (8.31) is applied to move the probe back inside the search space at (x_2, y_2). Here $F_{rep} = 0.85$.

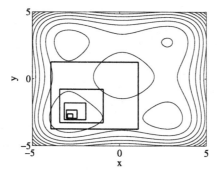

Figure 8.10 Contour plot of the Styblinski-Tang function. Plot showing the reduction in search area at iterations 20, 40, ...100.

In summary, the major steps in the CFO optimization are as follows:

BOX 8.4 Summary of CFO

Step 1: Generate an initial population of probes, using a systematic procedure, for example the placing equal number of probes in each dimension, distributed uniformly along the axes.
Step 2: Determine fitness of the function at each of population locations.
Step 3: Set the mass of each probe proportional to the fitness of the function at each probe location.
Step 4: Determine the acceleration of one probe towards another using (8.25).
Step 5: Determine the acceleration of each probe due to the effect of all the other probes (8.26).
Step 6: Determine the new locations of the probes due to the accelerations (8.27).
Step 7: If the required number of iterations is complete then end, else go to step 2.

Some modifications to the CFO have been proposed. Chen et al. (2016) have recognized the merits of CFO, particularly its easy implementation and the fact that it is

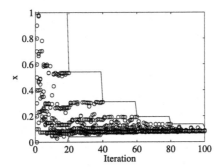

Figure 8.11 Optimization of function given by (8.34) with $F_{rep} = 0.5$. Plot showing the changes in the probe positions over 100 iterations. Also shown as a continuous line is the boundary of the search space that reduces at the 20th, 40th, 60th, etc. iteration.

deterministic. The rate of convergence of the CFO algorithm is based on the values of the parameters G, α and β. They claim that the convergence speed of the optimization process can be improved by introducing features of PSO and mutation. Liu and Tian (2015) have proposed a multi-start CFO. In the first stage, solutions are produced by two different initialization methods: probes distributed uniformly on each coordinate axis and probes distributed uniformly on the diagonals of the problem search space. In the second stage the initial solutions are improved by CFO.

8.11 SELECTED NUMERICAL STUDIES USING CFO

We now examine how the CFO algorithm performs on small number of test functions. It must be emphasized that this algorithm is distinct from all others described in this text since it is deterministic rather than stochastic. Since it is not stochastic, we have no need to perform repetitive runs to smooth out random effects, indeed for a given problem with a give number of probes in given initial locations, the results of multiple runs will be identical.

The first problem considered is the minimization of the one variable function

$$F(x) = \exp\left[-2\log_e(2)\left(\frac{x - x_0 - 0.08}{0.854}\right)^2\right]\sin^6\left[5\pi\left\{(x - x_0)^{0.75} - 0.05\right\}\right] \quad (8.34)$$

in the range 0 to 1. Here we take $x_0 = 0$.

To find the minimization of this function we will determine the maximum of $-F(x)$ using the CFO algorithm. Figure 8.11 and Figure 8.12 both show the initial positions of 11 probes, distributed at even intervals along the x axis at 0, 0.1, 0.2, ..., 1, and their subsequent values over 100 iterations. The only difference between the two optimizations is that the chosen value of F_{rep} in Figure 8.11 is 0.5 and in Figure 8.12 it is 0.95.

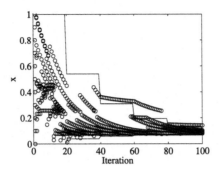

Figure 8.12 Optimization of function given by (8.34) with $F_{rep} = 0.95$. Plot showing the changes in the probe positions over 100 iterations. Also shown as a continuous line is the boundary of the search space that reduces at the 20th, 40th, 60th, etc. iteration.

After 100 iterations, most of the probes have clustered together at 0.079715 which is close to the minimum of $x = 0.0796875$.

It is seen that the two values of F_{rep} cause the positions of the probes to be quite different during the iterations although after approximately 100 iterations the probes converge to a value in the region of $x = 0.0796875$. Note that Figure 8.12 shows some probes lying outside of the reduced search area. For example from iteration 40 to iteration 79, one probe remains continuously outside of the search area. This unexpected result is explained as follows. For high values of F_{rep}, any probe trying to move outside the search area take a value very close its previous value, as can be seen from (8.32) and (8.33). Now at the 39th iteration, one probe has a value that is inside the search area, but at the next iteration (the 40th) the search area reduces. The probe which is outside the new search area takes a value close it its previous value, which had a value that is now outside of the search space. Subsequent probe positions are outside the search space and so take up a value close to the proceeding value which, of course, was outside the search space so the probe continues to be outside the search space. This, of course, does not happen when $F_{rep} = 0.5$.

We now consider the optimization of the Styblinski-Tang function in two variables. Because CFO determines the maximum of a function, we must optimize the negative of this function to find the maximum of the negative function, i.e. the minimum of the original function. Figure 8.13 shows the initial location of 12 probes (6 probes/dimension with $\gamma = 0.5$). We show the path of only one probe over 50 generations with $G = 2$, $\alpha = 2$, $\beta = 2$ and $F_{rep} = 0.5$. The figure shows that the probe is converging on the minimum of the function at approximately $x, y = -2.904$. Figure 8.14 shows the path of the probe in the region of this minimum. The probe moves in what superficially appears to be a series of random steps, but it is converging to a value close to the minimum of the Styblinski-Tang function. Of course, only one probe is shown, and there is no reason to assume that this probe gives the best solution.

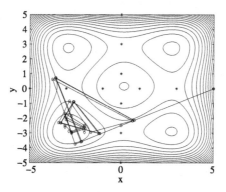

Figure 8.13 Optimization of the Styblinski-Tang function. The initial locations of the 12 probes are shown by stars. The path of only one probe, initially at (5, 0), is shown converging to the minimum of the function.

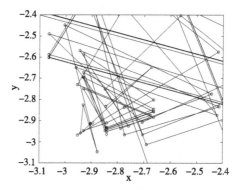

Figure 8.14 A close up of the region of the minimum of the Styblinski-Tang function showing path of only one probe, converging to the function minimum, $x, y = -2.904$.

Table 8.11 gives the results of optimizing the Styblinski-Tang function in 2 dimensions in the range $x = -5$ to 5, for 50 generations, with parameter values given in the title of the table. Note that for this algorithm n_{pop} is the number of probes used, so that n_{pop}/n_{var} is initial number of probes allocated to each dimension. Note that γ is defined in (8.26). For the 20 sets of initial conditions examined, a total of 360 probes have been "flown". For example, for all the cases in row 1 of Table 8.11 we have "flown" 6 probes per dimension i.e. 12 in total. The initial location of all the probes "flown" in this study are shown Figure 8.15.

For this problem and these parameters, Table 8.11 shows that the initial position of the probes is not significant, but this is not always the case. The results are generally good, particularly in view of the low number of generations used. All the results in the table provide a reasonable approximation to the global optimum although the best result is obtained when $\gamma = 0$.

Table 8.11 CFO: S-T2. $G = 2, \alpha = 2, \beta = 2, F_{rep} = 0.5, x = -5$ to 5

n_{gen}	n_{pop}/n_{var}	$\gamma = 0$	$\gamma = 0.25$	$\gamma = 0.50$	$\gamma = 0.75$	$\gamma = 1.00$
50	6	78.140724	78.331223	78.262716	78.223186	78.072581
50	8	78.292291	78.060839	78.327557	78.288272	78.266253
50	10	78.331574	78.243041	78.328864	78.298713	78.272852
50	12	78.134785	78.312901	78.323995	78.284916	78.267240
100	6	78.331741	78.331231	78.330596	78.327472	78.330191
100	8	78.329628	78.331195	78.330633	78.331229	78.324961
100	10	78.332073	78.327786	78.331678	78.330282	78.320847
100	12	78.332164	78.331812	78.331941	78.332203	78.331451

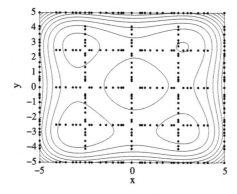

Figure 8.15 Initial location of the 360 probes used to generate Table 8.11.

Table 8.12 CFO: S-T2. $n_{gen} = 50, G = 2, \alpha = 2, \beta = 2, F_{rep} = 0.5, x = -5$ to $5, n_{pop}/n_{var} = 8$

	$\gamma = 0$	$\gamma = 0.25$	$\gamma = 0.50$	$\gamma = 0.75$	$\gamma = 1.00$
Parameters above	78.292291	78.060839	78.327557	78.288272	78.266253
$F_{rep} = 0.15$	77.560681	77.742574	78.288157	77.329808	77.462086
$F_{rep} = 0.95$	78.317753	78.218689	78.289081	78.114382	73.813640
$G = 5$	78.235484	78.290588	78.198357	78.245572	78.315168
$G = 0.5$	78.331307	78.306740	78.270846	78.261801	78.216135
$\alpha = 1$	78.301492	78.151485	77.984380	78.320077	78.300314
$\alpha = 3$	78.230142	78.153083	78.168638	78.007176	78.331775
$\beta = 1$	78.193724	78.249463	78.121559	78.283000	50.054287
$\beta = 3$	78.259086	78.232661	78.116141	78.192772	78.275531

Table 8.12 shows that the results of minimizing the Styblinski–Tang function are generally very good and not significantly influenced by the changes in the basic parameter values, except for $\gamma = 1$ with $F_{rep} = 0.95$ and also with $\beta = 1$. These two results are poor.

Table 8.13 CFO: ROS2, $n_{gen} = 50$, $G = 2$, $\alpha = 2$, $\beta = 2$, $F_{rep} = 0.5$, $x = -5$ to 5

n_{pop}/n_{var}	$\gamma = 0$	$\gamma = 0.25$	$\gamma = 0.50$	$\gamma = 0.75$	$\gamma = 1.00$
6	−0.004238	−0.205366	−0.025187	−0.760185	−0.009485
8	−0.079963	−0.448665	−1.000000	−0.033416	−0.995221
10	−0.000389	−0.036161	−0.015101	−0.075933	−0.000063
12	−0.000344	−0.388964	−0.034357	−0.021114	−0.002185

Table 8.14 CFO: RAS6, $G = 2$, $\alpha = 2$, $\beta = 2$, $F_{rep} = 0.5$, $x = -5$ to 5

n_{pop}/n_{var}	$\gamma = 0$	0.125	0.25	0.375	0.5	0.625	0.75	0.875	1.00
50 gens									
3	0	−0.7	−12.0	−3.3	0	−3.3	−12.0	−0.7	0
4	0	−1.7	−18.9	−5.0	0	−5.0	−19.0	−1.7	0
5	0	−21.0	−3.2	−1.5	0	−1.5	−3.2	−21.0	0
6	0	−6.2	−4.8	−6.5	0	−6.5	−4.8	−6.2	0
100 gens									
3	0	−0.5	−9.8	−2.9	0	−2.9	−9.8	−0.5	0
4	0	−1.4	−7.3	−0.3	0	−0.3	−7.3	−1.4	0
5	0	−3.3	−3.2	−1.0	0	−1.0	−3.2	−3.3	0
6	0	−6.0	−0.0	−6.5	0	−6.4	−0.0	−6.0	0
200 gens									
3	0	−0.0	−6.0	−1.7	0	−1.7	−6.0	−0.0	0
4	0	−1.0	−6.0	−0.0	0	−0.0	−6.0	−1.0	0
5	0	−1.0	−1.0	−1.0	0	−1.0	−1.0	−1.0	0
6	0	−6.0	−0.0	−6.0	0	−6.0	−0.0	−6.0	0

Table 8.13 shows that the results of minimizing the Rosenbrock function are generally very good, particularly considering the small number of iterations used. Of the 20 results, two are less accurate. Increasing the number of iterations used would improve the overall accuracy.

A more demanding test of the CFO algorithm is Rastrigin's function in 6 variables. Here again Table 8.14 shows that the results are generally good and are excellent for $\gamma = 0$, $\gamma = 0.5$ and $\gamma = 1$. In general the algorithm works well with only a small number of generations but sometimes less accurate results occur. Note that in Table 8.14 we only show one decimal place in the results.

Table 8.15 shows the effect of different choices of F_{rep}. In this example, we have chosen $\gamma = 0.46$ rather than 0.5, for example, because Table 8.14 shows that the for $\gamma = 0.5$ the result is 0 for all choices of the number of probes. For this more difficult problem in 6 variables, $F_{rep} = 0.95$ provides a superior set of results.

Table 8.16 shows the results of seeking the minimum of the Schwefel function in 30 dimensions or variables by finding the maximum of the negative of this function.

Table 8.15 CFO: RAS6, $n_{gen} = 200$, $G = 2$, $\alpha = 2$, $\beta = 2$, $\gamma = 0.46$, $x = -5$ to 5

n_{pop}/n_{var}	$F_{rep} = 0.5$	$F_{rep} = 0.95$
4	−3.8828007	−0.0256195
6	−4.9776682	−0.0003488
8	−6.9668431	−0.0000006
10	−8.9705875	−0.9960279
12	−10.9461212	−0.0000255

Table 8.16 CFO: Schwefel30, $G = 2$, $\alpha = 2$, $\beta = 2$, $F_{rep} = 0.5$, $x = -500$ to 500

n_{pop}/n_{var}	$\gamma = 0$	$\gamma = 0.50$	$\gamma = 1.00$
$n_{gen} = 8$			
4	−7146.3562	−3828.9304	−7.6305
6	−4681.3929	−9800.5276	−392.9986
8	−4942.8978	−0.3924	−1011.8700
10	−3722.8831	−6515.292	−444.6077
12	−4105.6190	−3772.7255	−22.3353
$n_{gen} = 50$			
4	−3928.0128	−3631.7820	−7.6305
6	−3633.6737	−9786.7859	−52.3088
8	−3569.4186	−0.3924	−17.1252
10	−3555.7298	−6515.0761	−11.8193
12	−3555.4228	−3660.7104	−22.3353

Schwefel function is described in Appendix A. Whilst the exact minimum of this function is zero, after only 8 iterations one probe has determined the minimum of the function with an error of only −0.3924; a truly remarkable result. For this result, the CFO algorithm determines $x_i = 420.665$, except for x_{28} which is equal to 420.306. The exact value of every element of **x** at this minimum is 420.969. Note also that after 50 iterations several probe distributions are now very close to the minimum.

8.12 SOME APPLICATIONS OF CENTRAL FORCE OPTIMIZATION

The developer of the CFO algorithm, Formato, illustrated the use of the algorithm in the optimum design of a 32-element linear array (Formato, 2007) and tested it on 23 antenna optimization benchmarks (Formato, 2010). The results were compared to published performance data for other optimization algorithms and the author states the CFO acquits itself quite well. Qubati et al. (2010) provide further examples of its effectiveness the CFO algorithm by applying it to pattern synthesis for linear and circular array antennas. A new selection scheme is introduced that enhances CFO's global search ability while maintaining its simplicity. The improved CFO algorithm is applied to the design of a circular array with very good results. The performance of CFO on the antenna benchmarks and the synthesis problems is compared to that of other evolutionary algorithms.

Asi and Dib (2010) applied CFO to the optimal design of multilayer microwave absorbers in a specific frequency range. Several numerical examples are presented, in which the CFO results are compared with those found by other evolutionary algorithms. The authors state that the CFO results are comparable to those found by the differential evolution algorithm and better than those found by particle swarm optimization (PSO) and gravitational search algorithm (GSA).

Finally, Central Force Optimization (CFO) has been applied to the problem of Inverse Transient Analysis (ITA). ITA is a powerful approach for leak detection and calibration of friction factors in pressurized pipes. Using this method, a transient flow is initiated and pressures are measured somewhere in the system. Then, a nonlinear programming problem with a least-squares criterion objective function is developed to minimize discrepancies between the measured and calculated pressures at measurement sites. Haghighi and Ramos (2012) have applied Central Force Optimization to the problem of ITA. The results are then discussed and compared to the previous studies. It is concluded that CFO is easy to implement, computationally efficient and has a remarkable performance in solving leak detection problem.

8.13 CENTRAL FORCE OPTIMIZATION COMPARED WITH GSA

Both the GSA and CFO are inspired by the way masses move in a gravitational field and superficially they might appear similar. However, they differ in several respects as follows:

- Most obviously and significantly, CFO is deterministic where as the GSA is stochastic.
- The positions of the initial distribution of probes are deterministic and generally evenly spaced in each dimension of the problem in CFO, whereas in the GSA the initial position of the agents chosen randomly.
- The parameter G is constant in CFO, whereas it reduces in the GSA as the number of iterations increases, given by (8.17).
- The masses are computed differently. In the GSA the masses are *related* to the fitness of the function by (8.14) and (8.15) whereas in CFO the masses are *directly proportional* to the fitness of the function at the location of each mass.
- The accelerations, velocities and displacements are computed differently.

8.14 SUMMARY

In this Chapter we have described four algorithms inspired by the laws of physics rather than biology. Of the four algorithms, Simulated Annealing is well established, where as the Big Bang-Big Crunch, the Gravitational Search Algorithm and Central Force Optimization are quite new. We have tried the algorithms on a range of test problems and, although variable in performance, generally they work quite efficiently.

8.15 PROBLEMS

8.1   Use Simulated Annealing to minimize the one variable function $f(x) = \sqrt{x}\sin(x)$ in the range 1 to 4. Take successive values of the temperature as $T = 1$ and $T = 0.1$. For each temperature perform only two iterations. For the purpose of this exercise take the initial random value of x as 0.25 and successive neighboring random values of x as 1.25, 2.5, 2.7 and 3.0. To decide on probabilistic steps you may use the values 0.32 and 0.95. [NB this step is not required on every occasion.]

8.2   We wish to find the minimum of the function $f(x, y) = \sin(x)\sqrt{x} + \sin(y)\sqrt{y}$ in the search range $0 \le x \le 10$ and $0 \le y \le 10$ using the BB–BC algorithm.
The initial (randomly) chosen positions of the six search agents are [8.1, 9.1], [1.3, 9.1], [6.3, 1.0], [2.8, 5.5], [9.6, 9.6], and [1.6, 9.7]. Assume $\alpha = 0.5$, and for simplicity $\beta = 0$, compute two iterations of the BB–BC algorithm to find the new positions of the agents. For each iteration compute the mean of the function at the locations of the agents.
At the start of the second iteration the normally distributed random numbers are [0.73, −0.06], [0.71, −0.21], [−0.12, 1.49], [1.41, 1.42], [0.67, −1.21], and [0.72, 1.63]. If any agent moves out of the search area simply take the x or y value of the agent outside the search area to be 5.

8.3   We wish to find the minimum value of $f(x, y) = \sqrt{(x)}\sin(x) + \sqrt{(y)}\sin(y)$ in the search range $x = 0$ to 10 and $y = 0$ to 10. Compute one iteration of the GSA algorithm with $G_0 = 2$ to find the new positions of the agents. The initial positions of the four agents are [8.15, 9.06], [1.27, 9.13], [6.32, 0.98] and [2.79, 5.47]. All the initial velocities are zero. For the purposes of these calculations, assume all random numbers equal 0.5.

8.4   We wish to find the maximum value of $f(x, y) = \sqrt{(x)}\sin(x) + \sqrt{(y)}\sin(y)$ in the search range $x = 0$ to 10 and $y = 0$ to 10. Compute one iteration of the CFO algorithm with $G = 2$, $\alpha = 2$ and $\beta = 2$ to find the new positions of the probes. The initial positions of the four probes are [1, 3], [1, 6], [6, 2] and [7, 8]. The initial velocities of the probes are zero. If it is required, use $F_{rep} = 0.5$.

CHAPTER 9

Integer, Constrained and Multi-Objective Optimization

9.1 INTEGER OPTIMIZATION

In some practical optimization problems only integer solutions are valid. For example, the number of people employed, the number of vehicles in a fleet, the number of plants operating must all be positive integers.

It is tempting to assume that the best integer solution is found by determining the non-integer solution and then taking the nearest integer value of the non-integer solution. However, the integer value closest to the non-integer solution is frequently not the optimum integer solution. This is illustrated by the graph shown in Figure 9.1. The minimum value of the function shown in the range 4 to 12 is $f(x) = -0.8022$ when $x = 6.6044$. If we round this value of x to 7 then $f(x) = -0.6730$. However, when $x = 10$, $f(x) = -0.6950$ and this is the required integer minimum solution.

The genetic algorithm lends itself very well to integer optimization. Suppose we seek an integer solution to a problem in the range 0 to 63. Then if each member of the population is limited to 5 bits it is impossible to determine a solution that is not an integer in the range 0 to 63. Integer solutions to optimization problems can also be used as an index to select data values from a table, see Section 2.10 of Chapter 2.

Many problems which naturally involve integer solutions such as machine and job scheduling problems and the traveling salesman problem (TSP) often involve binary integers. These problems are known to be extremely hard to solve. For a discussion of the TSP problem see Section 7.6 of Chapter 7.

9.2 CONSTRAINED OPTIMIZATION

We now consider non-linear optimization problems that are briefly introduced in Chapter 1. Here the objective function is non-linear and the constraints may be linear or non-linear. We give the general statement of a non-linear optimization problem in Chapter 1 and we repeat it here as a focus for our discussion, thus

$$\text{minimize } f(\mathbf{x})$$
$$\text{subject to } g_i(\mathbf{x}) \geq 0, \quad i = 1, 2, \ldots p \qquad (9.1)$$
$$h_j(\mathbf{x}) = 0, \quad j = 1, 2, \ldots q$$

Introduction to Nature-Inspired Optimization
DOI: 10.1016/B978-0-12-803636-5.00009-8

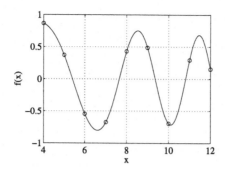

Figure 9.1 Plot of $f(x) = \exp(-x/30)\sin((x/3.05)^2)$ The minimum integer value is $x = 10$. The o denotes the integer values of the function.

where $\mathbf{x}$ is a vector of n elements. If any inequality constraints are of the form $g_i(\mathbf{x}) \leq 0$ then they may be converted to $g_i(\mathbf{x}) \geq 0$ by multiplying the constraint equations by -1. If the problem has no constraints it is called an unconstrained optimization problem.

There are several ways of dealing with constrained non-linear optimization problems; here we begin by describing one of the simplest methods. Suppose we seek to solve

$$\text{minimize } f(x, y)$$
$$\text{subject to } (x + y) \geq -2 \tag{9.2}$$

We will assume that we know that in the region of interest the function $f(x, y)$ has minimum values of less than 10. Let us seek the minimum value of the unconstrained function modified thus

$$g(x, y) = f(x, y)[(x + y) \geq -2] + 20[(x + y) < -2]$$

Here both $(x + y) \geq -2$ and $(x + y) < -2$ are defined to be equal to one if true and zero if false. Hence, when $(x + y) \geq -2$ is true and $(x + y) < -2$ is false, $g(x, y) = f(x, y)$, any minimum values found for g are, of course, those of f. Conversely, when $(x + y) < -2$ is true and $(x + y) \geq -2$ is false, $g(x, y) = 20$ and there are no minimum in this region. Thus, any algorithm being used to solve this problem will be prevented from finding a global minimum in this region. This is illustrated in Figure 9.2.

Another approach is to restrict the search area to the feasible region of the function. Any member of the trial population, agent or probe that moves in to the infeasible region is treated like a member of the population that has gone outside of the search space and is returned to the search space using a standard procedure appropriate to a particular algorithm. See for example, Section 8.10 of Chapter 8 which describes the way errant probes in Central Force Optimization are returned to the search space.

A formal way of solving the non-linear optimization problem is to use the Lagrange Multiplier Method. The method is not a purely numerical method, it requires the user

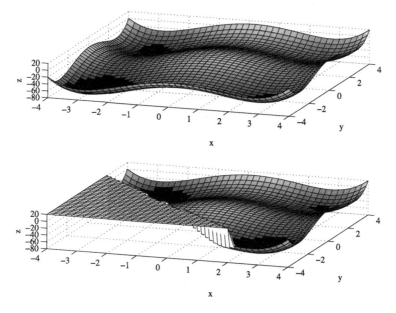

Figure 9.2 Two dimensional surface showing local maximum and minimum with 20 added as a constraint.

to apply calculus, and the resulting equations are normally nonlinear algebraic equations that must be solved numerically. For large problem this is impractical. However, it is theoretically important in the development of other, more practical, methods.

If constraints of the form of $g_j(\mathbf{x}) \geq b_j$ are present they must be converted to equalities by letting $\theta_j^2 = g_j(\mathbf{x}) - b_j$. If the constraint $g_j(\mathbf{x}) \geq b_j$ is violated, then θ_j^2 is negative and θ_j is imaginary. Thus, we have a requirement that for the constraints to be satisfied θ_j must be real. Thus the constraints become

$$\theta_j^2 - g_j(\mathbf{x}) + b_j = 0 \quad \text{where} \quad j = 1, 2, ..., q \tag{9.3}$$

These are equality constraints.

To solve (9.1) we begin by forming the expression

$$L(\mathbf{x}, \theta, \lambda) = f(\mathbf{x}) + \sum_{i=1}^{p} \lambda_i h_i(\mathbf{x}) + \sum_{j=1}^{q} \lambda_{p+j}[\theta_j^2 - g_j(\mathbf{x}) + b_j] = 0 \tag{9.4}$$

The function L is called the Lagrangian or the Lagrange function and the scalar quantities λ_i, where $i = 1, 2, ... p + q$ are called the Lagrange multipliers. We now minimize this function using calculus by setting the gradients of the function with respect to all

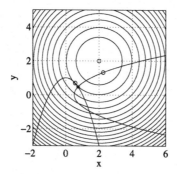

Figure 9.3 Plot of $f(x, y) = (x - 2)^2 + (y - 2)^2$ for various values of the objective function, and lines defining the constraint equations. '*' is the minimum value, 'o' are infeasible solutions and '□' is infeasible and is the unconstrained minimum of the function.

the variables to zero, thus:

$$\partial L/\partial x_k = 0, \quad k = 1, \dots n$$
$$\partial L/\partial \lambda_r = 0, \quad r = 1, \dots p + q$$
$$\partial L/\partial \theta_s = 0, \quad s = 1, \dots q$$

We will find that when we set the differentials with respect to λ_r to zero, we force both $h_i(\mathbf{x})$ ($i = 1, 2, \dots, n$) and $\theta_j^2 - g_j(\mathbf{x}) + b_j$ ($j = 1, 2, \dots, q$) to be zero because $\partial L/\partial \lambda_r$ is equal to $h_i(\mathbf{x})$ or $\theta_j^2 - g_j(\mathbf{x}) + b_j$. Thus the constraints are satisfied. If these terms are zero then minimizing (9.4) is equivalent to minimizing (9.1). If we are dealing with a quadratic function with linear constraints then the resulting equations are all linear and relatively easy to solve.

Consider the constrained optimization problem

$$\begin{aligned} \text{minimize } & f(x, y) = (x - 2)^2 + (y - 2)^2 \\ \text{subject to } & y^2 - x + 0.5 \leq 0 \\ & x^2 + y - 1 \leq 0 \end{aligned} \quad (9.5)$$

Figure 9.3 shows the problem graphically. The concentric circles centered on $x = 2$ and $y = 2$ are different values of $f(x, y)$, the function to be minimized. The parabolas are the equations of constraint. Thus the Lagrangian is

$$L = (x - 2)^2 + (y - 2)^2 + \lambda_1(\theta_1^2 + y^2 - x + 0.5) + \lambda_2(\theta_2^2 + x^2 + y - 1)$$

Taking partial derivatives of L with respect to each of the variables we have

$$\partial L/\partial x = 2(x - 2) - \lambda_1 + 2\lambda_2 x = 0$$

$$\partial L/\partial y = 2(y-2) + 2\lambda_1 y + \lambda_2 = 0$$
$$\partial L/\partial \lambda_1 = \theta_1^2 + y^2 - x + 0.5 = 0$$
$$\partial L/\partial \lambda_2 = \theta_2^2 + x^2 + y - 1 = 0$$
$$\partial L/\partial \theta_1 = 2\theta_1 \lambda_1 = 0$$
$$\partial L/\partial \theta_2 = 2\theta_2 \lambda_2 = 0$$

Even in this simple example, we are faced with the problem of solving six nonlinear simultaneous equations. However, we can deduce from the last two equations that we must solve the first four equations under four sets of conditions, thus:

Case 1 Assuming $\theta_1^2 = 0$ and $\theta_2^2 = 0$ and solving the first four equations we have $x = 0.7250$, $y = 0.4743$, $\lambda_1 = 2.3027$, $\lambda_2 = 0.7891$. The function value is $f = 3.9532$. This solution is feasible.

Case 2 Assuming $\lambda_1 = 0$ and $\theta_2^2 = 0$ and solving the first four equations gives $x = 0.5536$, $y = 0.6936$, $\theta_1^2 = -0.4274$, $\lambda_2 = 2.6128$. This is an infeasible solution.

Case 3 Assuming $\theta_1^2 = 0$ and $\lambda_2 = 0$ and solving the first four equations gives $x = 2.2549$, $y = 1.3247$, $\lambda_1 = 0.5097$ $\theta_2^2 = -5.4092$. This is an infeasible solution.

Case 4 Assuming $\lambda_1 = 0$ and $\lambda_2 = 0$ and solving the first equations gives $x = 2$, $y = 2$, $\theta_1^2 = -2.5$ $\theta_2^2 = -5$. The function value is $f = 0$. This is an infeasible solution and is actually the unconstrained minimum of the function.

Figure 9.3 shows the function, the constraints and the four solutions, only one of which is feasible and satisfies both constraints. This solution is the constrained minimum. Two solutions satisfy only one constraint each and under that condition, they give minimum values of the function. The fourth solution satisfies neither constraint and is the unconstrained minimum.

We now describe a technique that reduces the constrained optimization problem to an unconstrained one. It was developed by Fiacco and McCormick (1968, 1990) and is called the Sequential Unconstrained Minimization Technique (SUMT). The method converts the solution of a constrained optimization problem to the solution of a sequence of unconstrained problems.

Consider the optimization problem defined by (9.1). By using barrier and penalty functions the requirements of the constraints can be included with the function to be minimized so that the problem is converted to the unconstrained problem:

$$\text{Minimize } f(\mathbf{x}) - r_k \sum_{i=1}^{p} \log_e(g_i(\mathbf{x})) + \frac{1}{r_k} \sum_{j=1}^{q} h_j(\mathbf{x})^2 \tag{9.6}$$

Notice the effect of the added terms. The first term imposes a barrier at zero on the inequality constraints in that as the $g_i(\mathbf{x})$ approaches zero the function approaches minus infinity thus imposing a substantial penalty. The Figure 9.4 illustrates this. The last

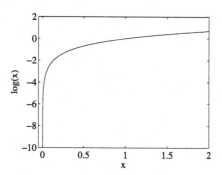

Figure 9.4 Graph of $\log_e(x)$.

term encourages the satisfaction of the equality constraints $h_j(\mathbf{x}) = 0$ since the smallest amount is added when all the constraints are zero, otherwise a substantial penalty is imposed. This means that this approach encourages the maintenance of the feasibility of the solution. However, we must start with an initial solution which is within the feasible region of the inequality constraints. These methods are called interior point methods.

A sequence of problems are generated by starting with an arbitrarily large value for r_0 and then using $r_{k+1} = r_k/c$ where $c > 1$ and solving the resulting sequence of unconstrained optimization problems. The unconstrained minimization steps may of course present formidable difficulties for some problems. A simple stopping criterion is to examine the difference between the value of $f(\mathbf{x})$ between successive unconstrained optimizations. If the difference is below a specified tolerance then stop the procedure.

There are various alternatives to this algorithm: for example a reciprocal barrier function can be used instead of the logarithmic function given above.

The barrier term can be replaced by a penalty function term of the form

$$\sum_{i=1}^{p} \max(0, g_i(\mathbf{x}))^2 \qquad (9.7)$$

Using (9.7), this term will add a substantial penalty if $g_i(\mathbf{x}) < 0$, otherwise no penalty is applied. This method has the advantage that feasibility is not required and is called an exterior point method.

The SUMT method has its disadvantages because it may lead to an ill-conditioned problem and a sequence of unconstrained problems must be solved.

Consider the non-linear constrained optimization problem defined by

$$\text{minimize } f(\mathbf{x}) = |x_1 - 3| + |x_2 - 2|$$
$$\text{subject to } g(\mathbf{x}) = x_1 - x_2^2 \geq 0 \qquad (9.8)$$
$$h(\mathbf{x}) = x_1^2 + x_2^2 - 1.2^2 = 0$$

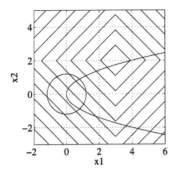

Figure 9.5 Plot of $f(\mathbf{x}) = |x_1 - 3| + |x_2 - 2|$ for various values of the objective function, and lines defining the constraint equation.

Table 9.1 Solution of the problem defined by (9.8) using the interior and exterior point methods

r	Interior	point		Exterior	point	
	x_1	x_2	$f(\mathbf{x})$	x_1	x_2	$f(\mathbf{x})$
10.00000	2.2500	2.5625	1.3125	1.3555	1.2246	2.4198
1.00000	1.2995	0.3238	3.3767	0.9248	0.9247	3.1505
0.10000	1.0141	0.6637	3.3222	0.8571	0.8571	3.2859
0.01000	0.8932	0.8031	3.3036	0.8494	0.8494	3.3012
0.00100	0.8563	0.8409	3.3028	0.8485	0.8487	3.3028
0.00010	0.8494	0.8476	3.3029	0.8486	0.8485	3.3029
0.00001	0.8486	0.8484	3.3029	0.8485	0.8486	3.3029

Figure 9.5 illustrates this problem. Notice that $h(\mathbf{x})$ is a circle of radius 1.2 and $g(\mathbf{x})$ is a parabola. The straight lines are values of the objective function to be minimized. (9.8) now becomes

$$\text{Minimize } |x_1 - 3| + |x_2 - 2| - r_k \log_e(x_1 - x_2^2) + \frac{1}{r_k}(x_1^2 + x_2^2 - 1.2^2)^2 \qquad (9.9)$$

For the exterior point method

$$\text{Minimize } |x_1 - 3| + |x_2 - 2| + \frac{1}{r_k}\min(0, (x_1 - x_2^2)) + \frac{1}{r_k}(x_1^2 + x_2^2 - 1.2^2)^2 \qquad (9.10)$$

Solving the equation for particular values of r_k is shown in Table 9.1 for both the interior and exterior methods. The exact optimum solution can be determined for this problem quite easily and is $x_1 = x_2 = 0.8485$ giving the minimum value of $f(\mathbf{x})$ equal to 3.3029. We see the table shows convergence to the correct solution.

If a classical optimization method is used, then the optimum computed solution $\mathbf{x}$ using say $r = 1$ is used as a starting value for the optimization using the next value of r, i.e. $r = 0.1$ in this case. However, using the methods of optimization described in this book, no initial values are required. The smaller the value of r, the more accurate the solution. However, very small values of r can lead to numerical instability.

Nature inspired optimization methods have been applied to a number of constrained optimization problems. Michalewicz (1995) provides a comprehensive review of methods to that date and discusses the different approaches for handling infeasible solutions.

Deb (2000) uses an approach which converts equality constraints to inequality constraints and then uses a penalty function to solve the constrained optimization problem by converting it to an unconstrained problem, as follows

$$F(\mathbf{x}) = f(\mathbf{x}) + \sum_{i=1}^{p} R_i(g_i(\mathbf{x}))^2 \tag{9.11}$$

In his paper Deb converts the equality constraints to inequalities as follows

$$g_{j+p}(\mathbf{x}) = \delta - |h_j(\mathbf{x})| \geq 0, \quad j = 1, 2, .., q$$

Consequently, (9.1) becomes

$$\text{minimize } f(\mathbf{x})$$
$$\text{subject to } g_i(\mathbf{x}) \geq 0, \quad i = 1, 2, ... p + q \tag{9.12}$$

Deb points out that there is considerable difficulty in choosing values for the parameters R_i of (9.11). If the constraints are normalized in some way, then the same value can be used for each R_i. After careful analysis and numerical experiments the Deb proposes that constrained problem becomes

$$F(\mathbf{x}) = \begin{cases} f(\mathbf{x}) \text{ if } g_k(\mathbf{x}) \geq 0, \quad k = 1, 2, ..., p + q \\ f_{max} + \sum_1^{p+q}(g_k(\mathbf{x})) \text{ otherwise} \end{cases} \tag{9.13}$$

where f_{max} is the objective function value of the worst feasible solution in the population. In his study, he solves nine constrained optimization problems using the Genetic Algorithm and reports good results.

Yeniay (2005) discusses how GAs may be applied to solve constrained optimization problems using penalty function methods. The author indicates that when using the GA, the exterior point penalty function is used in preference to interior point methods. Thus there is no need to start with a feasible initial population, because finding a feasible solution in many GA problems is a demanding problem in itself. The author describes a number of different penalty function methods and discusses the strength and weaknesses of these penalty function methods.

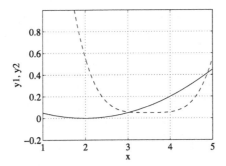

Figure 9.6 Plot of $y_1 = f_1(x)$ and $y_2 = f_2(x)$ showing the difficulty in choosing a compromise between minimizing one function or the other.

In a paper by Hu and Eberhart (2002) the authors discuss the application of the particle swarm optimization (PSO) method to constrained optimization problems and they describe two modifications which are applied to the original PSO algorithm. These are (i) during initialization all the particles are repeatedly initialized until they satisfy all the constraints and (ii) when calculating the local best and global best values only those particles in feasible space are counted. Hu and Eberhart tested the method successfully on 12 test functions.

9.3 INTRODUCTION TO MULTI-OBJECTIVE OPTIMIZATION

In some situations we wish to obtain simultaneously some sort of optimum solution for a number of objective functions, which may, or may not be subject to constraints or constrained only by the defined solution space. Clearly we should ask, is there a point which minimizes all the objective functions simultaneously? Since this is unlikely to be the case the best solution should be selected which is the best overall. This may be selected by a trade off between the objective function values, sometimes guided by the source or originator of the problem. To illustrate the difficulty of this problem, consider Figure 9.6. This figure shows two functions of $f_1(x) = 0.05(x-2)^2$ and $f_2(x) = 0.05 + 0.1(x - 3.5)^4$.

$$\text{rms error}(x) = (f_1(x) - 0)^2 + (f_2(x) - 0.05)^2$$

Note since the minimum of f_2 is 0.05 the error is $f_2(x) - 0.05$

$$e(x) = \max(|f_1(x) - 0|, |f_2(x) - 0.05|)$$
$$\text{minmax error} = \min(e(x))$$

The minimum of $f_1(x)$ is 0 at $x = 2$ but at $x = 2$, $f_2(x)$ is not a minimum. The minimum of $f_2(x)$ is 0.05 at $x = 3.5$. At $x = 3.5$ the rms and minimax errors are small.

Table 9.2 Values of $f_1(x)$ and $f_2(x)$

x	2	2.77	2.83	3.5
$f_1(x)$	0	0.0296	0.0344	0.1125
$f_2(x)$	0.5563	0.0784	0.0702	0.0500
rms error	0.5062	0.0411	0.0399	0.1125
minmax error	0.5062	0.0296	0.0344	0.1125

In contrast when $x = 2$ the rms and minimax errors are relatively large. When $x = 2.83$ (see Table 9.2) the rms error is a minimum and when $x = 2.77$ the minimax error is a minimum. Thus, there is no correct "best" solution, and it is up to the originator of the problem to choose which solution is the best for them. Clearly there is a range of values of x that might be chosen.

9.4 PARETO FRONT

The multi-objective optimization problem may be defined in terms of the objective functions that must be optimized. Thus if these are $f_1(\mathbf{x})$, $f_2(\mathbf{x})$, ..., $f_p(\mathbf{x})$, where $\mathbf{x}$ is a vector of variables. This may be defined as:

$$\min[f_1(\mathbf{x}), f_2(\mathbf{x}), ..., f_p(\mathbf{x})] \text{ where } \mathbf{x} \in \mathbf{S} \tag{9.14}$$

Here $\mathbf{S}$ is the feasible solution space and we have selected a minimization class of optimization problem.

To address the problem of the nature of the solutions of this type of problem which narrows down the solution possibilities, we introduce the concept of the Pareto Optimal solution. This is defined as a solution which cannot be improved upon without causing detriment to other solutions. For minimization problems this type of solution may be formally defined by considering two feasible solutions $\mathbf{x}^{(1)}$ and $\mathbf{x}^{(2)}$, then $\mathbf{x}^{(1)}$ dominates $\mathbf{x}^{(2)}$ if the equations 1 and 2 are both satisfied:

$$f_i(\mathbf{x}^{(1)}) \le f_i(\mathbf{x}^{(2)}) \text{ for } i = 1, 2, ..., p \tag{9.15}$$

$$f_k(\mathbf{x}^{(1)}) < f_k(\mathbf{x}^{(2)}) \text{ for at least one of the indexes } k = 1, 2, ..., p \tag{9.16}$$

Then for minimization problems the solution $\mathbf{x}^{(1)}$ is called Pareto optimal if there is no other solution which dominates it. However there may be many such solutions and this set of solutions is called the Pareto Front.

We now consider the task of determining the Pareto front for the two functions

$$f_1(x_1, x_2) = x_1^2 + x_2^2$$

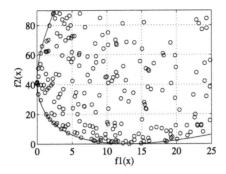

Figure 9.7 Plot of $f_1(x_1, x_2) = x_1^2 + x_2^2$ against $f_2(x_1, x_2) = 4(x_1 - 2)^2 + 3(x_2 - 3)^2$ showing the Pareto front and evaluations of the functions for randomly chosen values of x_1 and x_2.

and

$$f_2(x_1, x_2) = 4(x_1 - 2)^2 + 3(x_2 - 3)^2$$

We begin by defining the function

$$g(x_1, x_2) = (1 - \lambda)f_1(x_1, x_2) + \lambda f_2(x_1, x_2) \tag{9.17}$$

Thus

$$g(x_1, x_2) = (1 - \lambda)(x_1^2 + x_2^2) + \lambda(4(x_1 - 2)^2 + 3(x_2 - 3)^2)$$

and this simplifies to

$$g(x_1, x_2) = x_1^2(1 - 3\lambda) + x_2^2(1 - 2\lambda) - 16x_1\lambda - 18x_2\lambda + 43\lambda$$

To find the minimum of $g(x_1, x_2)$ we differentiate with respect to x_1 and x_2 and set these expressions to zero, thus

$$\partial g/\partial x_1 = 2x_1(1 + 3\lambda) - 16 = 0\lambda \tag{9.18}$$
$$\partial g/\partial x_2 = 2x_2(1 + 2\lambda) - 18 = 0\lambda \tag{9.19}$$

Thus $x_1 = 8\lambda/(1 + 3\lambda)$ and $x_2 = 9\lambda/(1 + 2\lambda)$. Clearly, we see that if $\lambda = 0$ the values of x_1 and x_2 minimize $f_1(x_1, x_2)$, and if $\lambda = 1$, the values of x_1 and x_2 minimize $f_2(x_1, x_2)$.

For specific values of λ we can derive values for x_1 and x_2, and substituting these values into f_1 and f_2 provides values of these functions for specific values of λ. Thus we can plot the Pareto front, as show in Figure 9.7. This figure also shows values of $f_1(\mathbf{x})$ and $f_2(\mathbf{x})$ plotted using random values of x_1 and x_2. We see that most points are suboptimal, but some points fall close to, or on, the Pareto front and are thus candidate for a possible best solution. If the Pareto front line had not been drawn, the location of the Pareto front could be deduced.

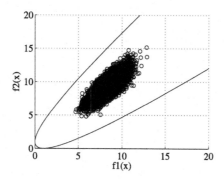

Figure 9.8 Plot of $f_1(\mathbf{x})$ against $f_2(\mathbf{x})$, where each function has 50 variables or dimensions. The plot shows the Pareto front and evaluations of $f_1(\mathbf{x})$ and $f_2(\mathbf{x})$ for 10,000 randomly chosen values of $\mathbf{x}$ where $\mathbf{x}$ is a trial vector of 50 elements.

It might seem that if we cannot derive an expression for the Pareto front, and in most situations it is not possible, we can approximate to it by evaluating random values of the functions. However, consider the even simpler functions

$$f_1(\mathbf{x}) = (1/n_{var}) \sum_{i=1}^{n_{var}} x_i^2$$

and

$$f_2(\mathbf{x}) = (1/n_{var}) \sum_{i=1}^{n_{var}} (x_i - 1)^2$$

Obviously $f_1(\mathbf{x})$ has a minimum value of zero when $x_i = 0, i = 1, ..., n_{var}$ and $f_2(\mathbf{x})$ has a minimum value of zero when $x_i = 1, i = 1, ..., n_{var}$.

Figure 9.8 provides a plot of f_1 against f_2 when $n_{var} = 50$ and the Pareto front is shown (since again it is easily computed). However, if we plot $10,000$ random evaluations of these functions, we see that none of the evaluations is close to optimal. Thus, we cannot obtain the Pareto front by random evaluations of the two, or more functions. A more efficient method is required. In this description of the Pareto front we have taken a simple approach using calculus. For more complex problems this may present difficulties.

9.5 METHODS FOR SOLVING THE MULTI-OBJECTIVE OPTIMIZATION PROBLEM

We now briefly describe a selection of methods for the optimizing multi-objective functions. Fuller descriptions for these methods are described in Deb (2001) and Narzizi (2008).

Weighted sum method. In this approach the multi-objective function problem is converted to minimizing a linear combination of the objective functions. Thus, we have

$$\min\{\lambda_1 f_1(x) + \lambda_2 f_2(x), \dots + \lambda_p f_p(x)\} \text{ where } x \in S \qquad (9.20)$$

In (9.20), $0 \leq \lambda_m \leq 1$ for $m = 1, 2, \dots, p$ and S is the constraint set.

However, the problem here is, how are the values of the scalars λ_m selected? The restriction $\sum \lambda_m = 1$ may also be imposed, but alternatives have been used. Although simple to use and guaranteeing the existence of a solution for some set of λ_m the method does not ensure a set of uniformly distributed Pareto solutions for uniformly distributed values of λ_m.

An important recent development using the weighted sum technique was described in Pereyra et al. (2013). In this paper they addressed the problem of solving a constrained two objective function optimization problem, but provided the means of producing an equi-spaced discrete approximation to the Pareto front. This is achieved by solving a sequence of problems which take the form:

$$\min_{(x,\lambda)} \quad (1 - \lambda)f_1(x) + \lambda f_2(x) \qquad (9.21)$$

$$\text{Subject to:} \quad x \in S, 0 \leq \lambda \leq 1$$

$$||f(x) - f_{prev}||^2 = \gamma^2$$

We note that λ is now an additional variable of the problem and initially $f_{prev} = f(x_0)$. Once this is solved, producing the solution (x_1, λ), then set $f_{prev} = f(x_1)$. This is the next problem in the sequence, producing the solution (x_2, λ) and so on. This will produce the set of solutions $(f(x_i), x_i)$ which is the discrete approximation to the Pareto front. Now a key point of this method is in the selection of γ. The value of this is defined as:

$$\gamma = \alpha ||f(x_0) - f(x_{(l+1)})||/l$$

Where x_0 and $x_{(l+1)}$ are the minimizers of $f_1(x)$ and $f_2(x)$ respectively implying l separate points between the end points. The constant $\alpha > 1$ is a user selected value. We can now see that the constraint $||f(x) - f_{prev}||^2 = \gamma^2$ imposes the required separation between the successive values of the function and consequently the points of the Pareto front.

Weighted metric methods. For these methods different combinations of the objective functions are used which take the form:

$$\min \left(\sum_{k=1}^{m} w_k |f_k(x) - z_k^*|^p \right)^{1/p} \quad \text{where } x \in S \qquad (9.22)$$

Where weights are non-negative, where p may take the value 1 or 2 for example, z_k^* is the ideal solution and S is the constraint set. This has the disadvantage that the ideal solution requires that each objective function is optimized. Note that (9.17) is a particular case of (9.22) for two functions.

Benson's method.   This avoids the estimation of the ideal solution by using z_k^0 as a value selected from the feasible region of the problem is taken and uses the new objective function:

$$\sum_{k=1}^{m} \max(0, (z_k^0 - f_k(x)) \text{ where } x \in S \tag{9.23}$$

$$\text{Subject to: } f_k(x) \le z_k^0 \text{ for all } k$$

Clearly this has the disadvantage of introducing additional constraints.

Lexicographic method.   An *a priori* method with known ranking. In practice it may often be the case that the objective functions can be characterized according to their importance, this ranking can be used to provide an approach to the multi-objective function problem. Thus assuming the vector $[f_1(x), f_2(x), ..., f_k(x)]$ is ranked in order the importance of the objective functions, then the sequence of single objective function problems are solved for $r = 1, 2, 3, ...$

$$\min f_r(x) \tag{9.24}$$

$$\text{Subject to } f_k(x) \le z_k^* \text{ where } k = 1, 2, ..., r-1$$

where z_1^* is the solution of the problem $\min(f_1(x))$ subject to $x \in S$. We note that one constraint is added for each problem we solve. Again additional constraints are added, but gradually.

There are many other techniques for multi-objective optimization; this is a brief introduction.

9.6 SUMMARY

In this Chapter we have described different classes of optimization problems, including integer programming, constrained optimization and the very difficult problem of multi-objective optimization. It must be emphasized that this is a brief introduction to this difficult area of optimization. A more advanced study is beyond the scope of this text.

CHAPTER 10

Recent Developments and Comparative Studies

10.1 INTRODUCTION

In this chapter we discuss briefly recently published papers, trends and developments. In addition we perform some comparisons of selected optimization algorithms and consider their performance on a range of test problems. The studies are carried out on standard test problems with a varying number of variables/dimensions and defined in varying sizes of solution space. Test problems also vary in difficulty in terms of the complexity of their multi-variable surface profile. For example functions may have deep shallow valleys, sharp or discontinuous changes in the function values, isolated, closely clustered or ill defined minima, ridges and plateaux. Thus additional numerical studies will be carried out on how algorithms cope with test problems having specific challenging features e.g. Easom's function and Rosenbrock's modified function.

These numerical studies are not meant to be definitive investigations but a limited illustrative selection of those that can be used to assess the behavior of algorithms. Suggestions for further investigative studies of the selected algorithms which readers can undertake will be given.

10.2 OTHER NATURE INSPIRED OPTIMIZATION ALGORITHMS

Although we have described in detail several nature inspired algorithms in Chapters 2 to 8, the field is very broad and here we briefly mention some interesting algorithms that we have not previously described.

The Strawberry optimization algorithm was introduced by Salhi and Fraga (2011) and is based on the remarkably efficient manner in which strawberry plants propagate and sustain themselves. These plants send out runners to propagate and sustain themselves in a specific way that can be related to global optimization. Runners are sent out in all directions but not in a purely random way. If the plant discovers an area with good soil and light then the plant settles that area by sending out relatively short runners so that other plants can benefit from the local circumstances. However, if the area is not propitious, longer runners are sent out to discover new areas where the plants can prosper. The length of these runners is inversely proportional to the fitness of the plants. After the formal mathematical specification of the rules this leads to an algorithm which is effective for solving non–linear optimization problems and is capable of finding global optima. The literature confirms that this algorithm has had considerable success for a range of demanding problems.

The Bat algorithm is based on the bats use of echolocation to efficiently find sources of food and roosting locations and was introduced by Yang (2010). All bats use echolocation to determine the location of their objectives and can distinguish these from background noise. Bats fly randomly and by adjusting the frequency and rate of the emitted pulses, dependent on their closeness to the target, in order to improve the efficiency of the search process. The values of loudness and pulse rate have to be adjusted consistent with the loudness decreasing and the pulse rate increasing as the prey is found. These are the main features on which the algorithm is based. The literature reports that this algorithm has had considerable success for a range of demanding problems.

The Krill herd optimization is a swarm intelligence method introduced by Gandomi and Alavi (2012), based on herding behavior of ocean krill and their search for food. Clearly if a path can be found which minimizes the distance traveled by the Krill herd as a whole to a good food source it will greatly improve their survival chances. This algorithm is interesting in that involves the use of aspects of GAs and a modified hybrid form has been proposed which uses simulated annealing. The fitness of each Krill depends on the distance from food and the distance from the point of highest density. Krill individuals try to maintain high density to the benefit of the whole herd. A comprehensive review of the Krill algorithm is provided by Bolaji et al. (2016).

The progress of rivers from start to destination often proceeds by a complex pattern of twists and turns as it selects a particular route. This process of path finding has inspired an alternative particle swarm algorithm based on intelligent water drops, by considering the river to be composed by many water drops that interact with each other and the soil of the river bed and banks to produce the particular route. The algorithm is based on the water velocity and soil interaction. Clearly the water drop velocity can change as the river progresses and in addition the motion of the drops may remove some soil as it passes. The velocity of the intelligent water drops is a function of the inverse of the soil interaction between two locations. The intelligent water drops gathers soil as it proceeds. The amount of soil added to the intelligent water drops is non-linearly proportional to the inverse of the time to pass between locations assuming the passage of the intelligent water drops is incremental. Using these principles the Intelligent Water Drops (IWD) algorithm was developed by Shah-Hosseini (2008).

The Colliding Bodies Optimization (CBO) is a metaheuristic optimization algorithm developed by Kaveh and Mahdavi (2014) and is loosely based on the behavior of colliding bodies. When two bodies collide the principle of conservation of momentum determines their velocities after the collision. In the algorithm each agent or body has a mass related to the fitness of its current position. After a collision the two bodies separate with new velocities and their displacement after collision is related to these velocities. This collision causes the agents to move toward better positions in the search space.

CBO is a simple formulation and finds optimum of functions and the authors state that the method does not depend on any internal parameter. Numerical results show that CBO is competitive with other metaheuristics.

This is far from an exhaustive list of nature inspired optimization algorithms. de Castro (2007), Biswas et al. (2013) and Fister et al. (2013) all review nature inspired algorithms for optimization and Siddique and Adeli (2015) provide a useful review of methods and suggest future directions for nature inspired computing.

In recent years there has been an explosion of the range and variety of NIO methods. It has been suggested that some methods lack novelty but this is obscured by the various different analogies to biological and physical phenomena. The harmony search algorithm (not described in this text) is inspired by the improvisation process of jazz musicians but Weyland (2010) stated that the harmonic search is just a special case of evolution strategies, although it was not suggested that this was evident to the authors of the harmony search algorithm. One of the authors of the harmony search algorithm published a rebuttal (Geem, 2010). Probably approaching 100 methods have been devised and published and this does raise important issues as to which are the most effective methods overall, which are the most effective for specific problems, and which provide a distinctive and innovative approach to the solution of problem.

10.3 COMPARATIVE STUDIES OF SELECTED METHODS ON SPECIFIC TEST PROBLEMS

Before embarking on comparative studies it is important to highlight how difficult a fair and equitable comparison of different methods is, particularly in regard to the intelligent selection of the key parameters for each algorithm and differences in programming style. We refer the reader to Chapter 1 where these matters have been discussed. In addition we must take the average of a number of runs to mitigate untypical random effects. The general basis of comparison we have selected is in terms of the number of iterations or generations required to obtain the same specific accuracy for each algorithm. Assessment of results may also be in terms of time/function evaluations. We discuss this further after the results have been obtained.

We have decided on the comparative study of a selection of the methods described in the book. These are: the Artificial Bee Colony (ABC) algorithm, the Differential Evolution (DE) algorithm, and Cuckoo Search algorithm (CSA), and in two studies, the Firefly algorithm (FFA). We consider these tests may be more revealing if separate comparisons are done for the effects of some of the key features of global optimization problems. These are increasing dimensionality or number of variables, the size of the solution space and the complexity of the test function surface. Consequently we have separated these studies and have provided a series of tables giving the numerical results for these aspects of test problems studied.

Table 10.1 Number of iterations required to minimize Rastrigin's function with n_{var} variables

n_{var}	ABC	DE	CS
10	431	967	1618
15	653	3692	2460
20	965	12917	3028
30	1581	NC	4656
40	2255	NC	6793

Table 10.2 Number of iterations required to minimize Rosenbrock's function with n_{var} variables

n_{var}	ABC	DE	CS
10	1907	3989	7241
15	2675	8434	8969
20	11411	16374	12991
30	18696	NC	31470
40	26896	NC	NC

In Table 10.1 we give the results of minimizing the Rastrigin function, a function with many optima, for different numbers of variables. The number of variables is designated in the table by the variable n_{var}. See Appendix A for a three dimensional plot of this function. The comparison is done between the ABC algorithm, the DE algorithm and CSA. These three algorithms are competitive for large dimension problems. The entries in the table provide the average results for 10 runs. The maximum number of iterations used is 40,000.

In Table 10.1 the entries give the number of iterations to achieve a reduction of the function value to less than 0.0005. The entry NC indicates the convergence criteria is not satisfied by the given maximum number of generations. The results show that the ABC algorithm provides the best results in terms of iterations by solving all the problems with the least number of iterations. The CS method also solves all the problems but takes more iterations. The DE method appears to have difficulty with the higher dimensional problems. Care should be taken with this assessment since the computational complexity of each iteration should also be considered.

We now repeat this experiment for a different function. In Table 10.2 the entries give the number of generations/iterations to achieve a reduction of the function value to less than 0.005 but this time using Rosenbrock's function which has a single minimum but in a shallow valley with 10 runs and a maximum of 40,000 iterations. See Appendix A for a plot of this function with two variables.

Table 10.3 Number of iterations required to minimize Easom's function with 2 variables for different size solution spaces

Range	ABC	DE	CS	FF
R1	22	61	42	47
R2	126	104	91	221
R3	1153	398	295	4245
R4	3421	823	991	NC

The entry NC indicates the algorithm does not converge to the required accuracy after the maximum number of generations. The results show that the ABC algorithm provides the best results but the number of iterations are far greater for this function for all the algorithms tested and the CS algorithm fails to converge for the 40 variable case.

We now consider tests on a function which presents specific difficulties. Easom's function is one which has a single sharply defined minimum in a relatively large solution space, see Appendix A for a three dimensional plot of this function. We consider the results of minimizing Easom's function over an increasingly large solution space determined by the ranges of values for the individual variables. This function is one with a single, very sharply defined, minimum at (π, π) with the value of the objective function equal to -1 and consequently there are only rare indications of a good search direction for any algorithm used for its solution. In the following tests we include the Firefly algorithm in the tests to see how it compares for this type of problem.

Table 10.3 gives the results of optimizing the two variable Easom's function within an increasingly large solution space denoted by Range, where R1 denotes the range for all variables $[-5, 5]$, R2 denotes $[-50, 50]$, R3 denotes $[-250, 250]$ and R4 denotes $[-500, 500]$ the entries give the number of iterations to achieve a reduction of the function to less than 10^{-5}. The average number of generations/iterations to obtain a specific accuracy for 10 runs of the algorithms are given. Note NC indicates incomplete convergence where the runs require consistently over 5,000 iterations.

The results of Table 10.3 show that all algorithms are successful in solving Easom's function except the Firefly algorithm in one case. However, the difficulty of the problem increases markedly with the size of the solution space. The DE and CS algorithms perform similarly and provide the best performance the ABC method in general requires many more iterations than the CS and DE. The Firefly algorithm is the least successful.

Table 10.4 shows the performance of algorithms on a set of functions having specific surface features. These were selected from the comprehensive list given by Jamil and Yang (2013). The ones selected are the Bird function, the Chichnadze function, the Table 2 or Holder function, the Alpine 2 function, the Griewank function and Rosenbrock's modified function. The three dimensional graphs of these functions are supplied in Appendix A and these give an indication of the complexity of the surface

Table 10.4 Number of iterations required to minimize specific functions

Range	ABC	DE	CS	FF
f_1	50	116	124	375
f_2	38	72	573	654
f_3	88	278	1255	1553
f_4	26	342	93	360
f_5	23	27	86	177
f_6	NC	NC	NC	NC

and any special features. The difficulty of many test functions has been examined in the Infinity77 study (currently available on the internet) and each function is given a rating between 0 and 100 which indicates the percentage of times the function was successfully solved by a range of algorithms. In the Infinity77 hardness table the Rosenbrock's modified function was one of the hardest receiving a score of only 8.42, the Griewank function was also classified as very hard receiving a score of 6.08. The Chichnadze function is an easier function but well in the more difficult half of the problems with a score of 40.33. The other functions are in the easier category of challenges. For each test function 10 runs were performed and a maximum of 2,000 iterations allowed and an accuracy of 0.0005.

In Table 10.4 f_1 denotes the Bird function, f_2 the Chichnadze function, f_3 the Griewank function, f_4 the Table 2 (Holder) function, f_5 the Alpine 2 function and f_6 is Rosenbrock's modified function. The results for the algorithms considered appear quite promising. In particular the ABC and DE methods are very successful in optimizing five of the functions, the ABC method being the best of the two in terms of the number of iterations required. The cuckoo search algorithm is also successful but does need significantly more iterations than both the ABC and the DE method. Finally for these problems the firefly method although optimizing five of the problems takes more iterations than the other methods. This version of the firefly algorithm used the normal distribution to generate random values.

None of the methods were wholly successful in solving Rosenbrock's modified function. However the methods did optimize the function for some runs and appear to oscillate between two alternative solutions, one of which is the global optimum. Consequently we give the number of times each of the optimization methods were successful in finding the global optimum to the same required accuracy in 20 runs, where 2000 generations were used. This provides a better insight into the behavior of the algorithms for this problem (see Table 10.5).

Clearly Rosenbrock's modified function is a challenge to all of the algorithms since the algorithms failed in the majority of cases to meet the convergence requirements. Although the Artificial Bee Colony algorithm was the best, the difference between that

Table 10.5 Number of successful minimizations of Rosenbrock's modified function

Algorithm:	ABC	DE	CS	FF
Succeeds	5/20	4/20	4/20	1/20

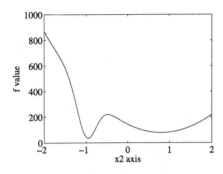

Figure 10.1 Showing a profile of Rosenbrock's modified function using $x_1 = -0.9$ and x_2 varies between -2 and 2.

algorithm, Differential Evolution and Cuckoo Search algorithm is negligible. In this study, the Firefly algorithm provided the least satisfactory performance. In fairness to the Firefly algorithm, it should be noted that recent developments of the algorithm are reported to improve its performance and merit further study.

To further investigate the reasons for the difficulty of Rosenbrock's modified function, we give a graphical illustration of the nature of the function by drawing a profile of the function taking x_1 as -0.9. Clearly the profiles will vary dependent on the profile taken, e.g. the shallow valley may be more shallow in some sections. This is illustrated by Figure 10.1.

This profile of the function helps to provide an explanation of the unusual difficulty of this function. It shows a shallow valley where a local minimum lies but also a steeper and deeper valley where the true minimum lies at $[-0.9, -0.95]$. Because of the shallowness of one of the valleys this requires significant exploration with little indication of good directions of search. In addition there is a barrier that the specific algorithm must surmount before the deeper valley where the global minimum lies can be explored (see Figure 10.2).

Finally we must emphasize that for all these studies the comparison between the algorithms could also be based on time or function evaluations, since each iteration will require a specific number of function evaluations and this may vary between algorithms. Time values depend on the specific machine on which the runs are performed.

The MATLAB profile function can be used to provide a detailed analysis of the algorithms that includes all these comparative factors; i.e., time and function evaluations and includes a detailed analysis of the process.

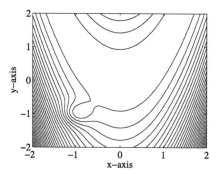

Figure 10.2 Figure shows the contour plot of Rosenbrock's modified function showing the global minimum at [−0.9, −0.95], but showing the part of the shallow region in which the algorithm must also search.

10.4 SUGGESTIONS FOR FURTHER STUDIES

Some studies apply to all algorithms because they consider a key common feature. Each algorithm involves deciding on a specific population or swarm size. In this book the population size is often taken as 30 but a larger or smaller size may be used. This is a factor which is often problem dependent larger populations being used by researchers for problems with large numbers of variables, complex solution spaces or large solution spaces. Thus studies on optimizing problems with 20 to 40 variables may benefit from large populations in the range 30 to 60 or even greater could prove useful.

When solving problems that are challenging, it is often necessary to increase the number of iterations or generations needed to solve the problem. Formula of the type an_{var} or an_{var}^{p}, where p is an integer, could be studied to see how effective they are in selecting the number of iterations.

Another group of studies would be to establish which parameters, if any, for a specific algorithm, are critical. For example in the cuckoo search algorithm deciding a good value for the parameter which decides how many nests should be abandoned may be critical for some problems. In the Differential Evolution the parameter F, which controls the amplitude of the differential evolution, and c, the cross over rate, must be carefully set in a specific range. Determining which are the best values for a range of test problems is a valuable study for this algorithm.

More demanding studies could be based on the suggested improvements or variations to specific algorithms suggested by researchers in the field. For example, in the ABC algorithm different formulae have been suggested for deciding on the selection of a promising nectar source, studies would be useful in deciding which is the best and in what circumstances. Other suggestions have been made in regard to the best way to generate the initial population for the specific algorithm, see Gao and Liu (2011). Since many algorithms require the generation of an initial population, this approach could be applied to many of them, replacing the standard method using a uniform random generator. For example, by using the method of Gao and Lui instead and comparing the results. In addition in searching the solution space, which is again a requirement of most nature based algorithms, alternative probability distributions have been used, e.g. the Lévy distribution, the normal distribution, the uniform distribution, the use of chaotic maps and other distributions. This approach was discussed in Fister et al. (2014) in relation to the Firefly algorithm. These distributions could be applied to other appropriate algorithms at the exploratory search stage to study how effective these alternative methods would be.

10.5 SUMMARY

In this concluding chapter we briefly described a small but important selection of recently developed optimization algorithms inspired by nature. These algorithms offer opportunities for further study and experimentation. This is a selection of a very large and growing number of nature inspired algorithms. We have performed basic illustrative numerical studied of the comparative performance of some algorithms described in this text. The results of these tests have been obtained in specific circumstances and we do not claim the results to be definitive.

10.6 PROBLEMS

The specific problems outlined here may be solved by the application of the ABC algorithm, DE algorithm and Cuckoo Search algorithm, using the MATLAB scripts supplied in Appendix B or any other available and appropriate software. It is expected that those readers who choose to use the MATLAB scripts provided have sufficient knowledge of MATLAB. To help any reader using these scripts, we also provide MATLAB functions for the Rosenbrock, Rastrigin and Egg Crate functions. No solutions are given; the solutions will not be identical due to the random nature of the processes.

10.1 Study the behavior of the cuckoo search algorithm in optimizing the egg crate function as the dimensions of the problem increases. Test the cuckoo search algorithm for the egg crate problem using 5, 10, 15, 20 and 25 variables. Perform 10 runs of the algorithm for each problem.

10.2 Establish if there is any discernible relationship between the number of variables in the egg-crate problem when using the Cuckoo search algorithm and the accuracy of the result, using the data from Problem 10.1.

10.3 Study the behavior of the cuckoo search algorithm in optimizing Rosenbrock's function as the dimensions of the problem increases. Test the cuckoo search algorithm for the egg crate problem using 5, 10, 15, 20 and 25 variables. Perform 10 runs of the algorithm for each problem.

10.4   For the ABC algorithm test the performance of the algorithm on the Rastrigin function with 10, 20 and 30 variables using 4000, 6000 and 8000 iterations respectively. Perform 10 runs of the algorithm.

10.5   For the ABC algorithm test the performance of the algorithm on the egg crate function with 10 variables using 4000 iterations and perform 20 runs of the algorithm.

10.6   For the ABC algorithm test the performance of the algorithm on the Rastrigin function with 30 variables using 20,000 iterations and perform 20 runs of the algorithm.

10.7   For the ABC algorithm test the performance of the algorithm on the Rosenbrock function with 10, 15 and 20 variables using 4000, 6000 and 8000 iterations respectively. Perform 10 runs of the algorithm.

10.8   For the Differential Evolution algorithm study the algorithms performance for solving Rastrigin's function with 12 variables using values of the F parameter equal to 0.5, 1, 1.5 and 2. Use 20 runs. Record your results and comment on their significance.

10.9   For the Differential Evolution algorithm study the algorithms performance for solving Rastrigin's function with 12 variables using values of the c parameter equal to 0.1, 0.4, 0.75 and 1. Use 20 runs. Record your results and comment on their significance.

10.10 For the Differential Evolution algorithm study the algorithms performance for solving Rosenbrock's function with 12 variables using values of the c parameter equal to 0.1, 0.4, 0.75 and 1. Use 20 runs. Record your results and comment on their significance.

10.11 For the Differential Evolution algorithm study the algorithms performance for solving the egg crate function with 12 variables using values of the c parameter equal to 0.1, 0.4, 0.75 and 1. Use 20 runs. Record your results and comment on their significance.

APPENDIX A

Test Functions

A.1 INTRODUCTION

A large number of functions have been used to test or benchmark optimization techniques and Jamil and Yang (2013) list 175 functions in a comprehensive literature review of benchmark functions, but even this collection is not exhaustive. Surjanovic and Bingham (2013) maintain a virtual library of simulation experiments on the web. Adoria and Diliman (2005) list 54 optimization functions in C code.

A.2 TEST FUNCTIONS

In this section are listed the test functions used in this text.

Ackley's Function (ACK) This function, in n_{var} variables, is defined by:

$$f(\mathbf{x}) = -20 \exp\left(-0.2\sqrt{\frac{1}{n_{var}}\sum_{i=1}^{n_{var}} x_i^2}\right) - \exp\left(\frac{1}{n_{var}}\sum_{i=1}^{n_{var}} \cos(2\pi x_i)\right) + 20 + e$$

The global minimum of the function is at $\mathbf{x}^* = [0, 0\ldots, 0]$ and $f(\mathbf{x}^*) = 0$. The function is usually searched in the hypercube $\mathbf{x} \in [-35, 35]$. A plot of this function in two variables is shown in Figure A.1.

Alpine 2 Function (ALP) This function, in n_{var} variables, is defined by:

$$f(\mathbf{x}) = \prod_{i=1}^{n_{var}} \sqrt{x_i}\sin(x_i)$$

The global minimum of the function is at $\mathbf{x}^* = [7.917.., 7.917.\ldots., 7.917..]$ and $f(\mathbf{x}^*) = 2.808^{n_{var}}$. The function is usually searched in the hypercube $\mathbf{x} \in [0, 10]$. A plot of this function in two variables is shown in Figure A.2.

Bird's Function (BIR) This function, in 2 variables, is defined by:

$$f(\mathbf{x}) = \sin(x_1)e^{(1-\cos(x_2))^2} + \cos(x_2)e^{1-\sin(x_1))^2} + (x_1 - x_2)^2$$

Introduction to Nature-Inspired Optimization
DOI: 10.1016/B978-0-12-803636-5.00020-7

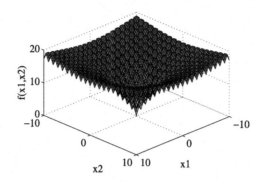

Figure A.1 Ackley's function with two variables.

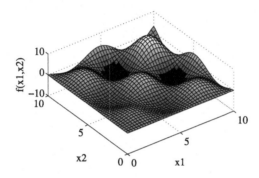

Figure A.2 Alpine 2 function with two variables.

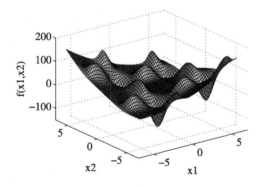

Figure A.3 Bird's function with two variables.

The global minimum of the function is at $\mathbf{x}^* = [4.70104,\ 3.15294]$ and $[-1.58214,\ -3.13024]$. Minimum is $f(\mathbf{x}^*) = -106.764537$. The function is usually searched in the area $\mathbf{x} \in [-2\pi,\ 2\pi]$. A plot of this function is shown in Figure A.3.

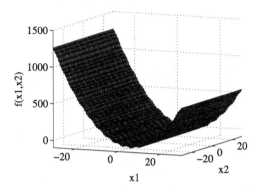

Figure A.4 Chinchnadze's function with two variables.

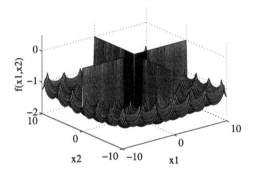

Figure A.5 Cross-in-Tray function.

Chinchnadze's Function   (CHI) This function, in 2 variables, is defined by:

$$f(\mathbf{x}) = x_1^2 - 12x_1 + 11 + 10\cos(\pi x_1/2) + 8\sin(5\pi x_1/2) - (1/5)^{0.5}\exp(-0.5(x_2 - 0.5)^2)$$

The global minimum of the function is at $\mathbf{x}^* = [5.90133,\ 0.5]$ and $f(\mathbf{x}^*) = -43.3159$. The function is usually searched in the hypercube $\mathbf{x} \in [-30,\ 30]$. A plot of this function in two variables is shown in Figure A.4.

Cross-in-Tray Function   [CRO] This function, in 2 variables, is defined by:

$$f(\mathbf{x}) = -0.0001\left(\left|\sin(x_1)\sin(x_2)\exp\left(\left|100 - \frac{1}{\pi}\sqrt{x_1^2 + x_2^2}\right|\right)\right| + 1\right)^{0.1}$$

The four global minimum of the function is at

$$\mathbf{x}^* = [\pm 1.3494066865 \quad \pm 1.3494066865] \text{ and } f(\mathbf{x}^*) = -2.06261218.$$

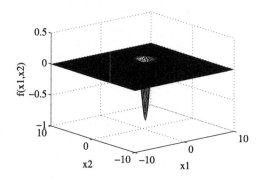

Figure A.6 Easom's function with two variables.

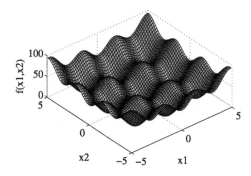

Figure A.7 Egg Crate function with two variables.

The function is usually searched in the region $\mathbf{x} \in [-10, 10]$. A plot of this function is shown in Figure A.5.

Easom's Function   [EAS] This function, in 2 variables, is defined by:

$$f(\mathbf{x}) = -\cos(x_1)\cos(x_2)\exp\left(-(x_1 - \pi)^2 - (x_2 - \pi)^2\right)$$

The global minimum of this function is at $\mathbf{x}^* = [\pi. \quad \pi]$ and $f(\mathbf{x}^*) = -1$. The function is usually searched in the region $\mathbf{x} \in [-100, 100]$. A plot of this function is shown in Figure A.6.

Egg Crate Function   [EGC] This function, in 2 variables, is defined by:

$$f(\mathbf{x}) = x_1^2 + x_2^2 + 25(\sin^2(x_1) + \sin^2(x_2))$$

The global minimum of this function is at $\mathbf{x}^* = [0, 0]$ and $f(\mathbf{x}^*) = 0$. The function is normally searched in the region $\mathbf{x} \in [-5, 5]$. A plot of this function is shown in Figure A.7.

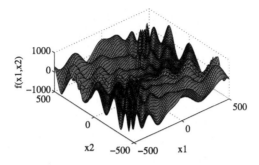

Figure A.8 Egg Holder function with two variables.

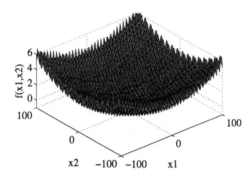

Figure A.9 Griewank's function with two variables.

Egg Holder Function   [EGH] This function, in n_{var} variables, is defined by:

$$f(\mathbf{x}) = \sum_{i=1}^{n_{var}-1} \left[-(x_{i+1} + 47) \sin\left(\sqrt{|x_{i+1} + x_i/2 + 47|}\right) - x_i \sin\left(\sqrt{|x_i - (x_{i+1} + 47)|}\right) \right]$$

The global minimum of the function is at $\mathbf{x}^* = [512, 404.2319]$ and has a value of $f(\mathbf{x}^*) = 959.964$. The function is usually searched in the hypercube $\mathbf{x} \in [-512, 512]$. A plot of this function is shown in Figure A.8.

Griewank's Function   [GRI] This function, in n_{var} variables, is defined by:

$$f(\mathbf{x}) = \sum_{i=1}^{n_{var}} \frac{x_i^2}{4000} - \prod_{i=1}^{n_{var}} \cos\left(\frac{x_i}{\sqrt{i}}\right) + 1$$

The global minimum of the function is at $\mathbf{x}^* = [0, 0]$ and $f(\mathbf{x}^*) = 0$. The function is usually searched in the hypercube $\mathbf{x} \in [-100, 100]$. A plot of this function in two variables is shown in Figure A.9.

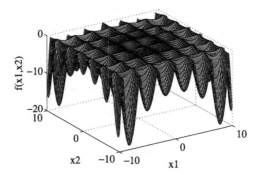

Figure A.10 Holder's table function.

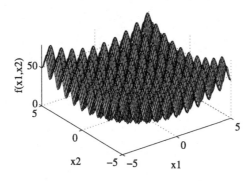

Figure A.11 Rastrigin's function with two variables.

Holder Table Function   [HOL] This function, in 2 variables, is defined by:

$$f(\mathbf{x}) = -\left| \sin(x_1)\cos(x_2)\exp\left(\left|1 - \frac{1}{\pi}\sqrt{x_1^2 + x_2^2}\right|\right)\right|$$

The global minimum of the function is $f(\mathbf{x}^*) = -19.2085$ at $\mathbf{x}^* = [8.05502,$ $9.66459]$, $[8.05502, -9.66459]$, $[-8.05502, 9.66459]$, $[-8.05502, -9.66459]$. The function is usually searched in the square $\mathbf{x} \in [-10, 10]$. A plot of this function is shown in Figure A.10.

Rastrigin's Function   [RAS] This function, in n_{var} variables, is defined by:

$$f(\mathbf{x}) = 10n_{var} + \sum_{i=1}^{v_{var}} \left(x_i^2 - 10\cos(2\pi x_i)\right)$$

The global minimum of the function is at $\mathbf{x}^* = [0 \quad 0\ldots0]$ and $f(\mathbf{x}^*) = 0$. The function is usually searched in the hypercube $\mathbf{x} \in [-5.12, 5.12]$. A plot of this function in two variables is shown in Figure A.11.

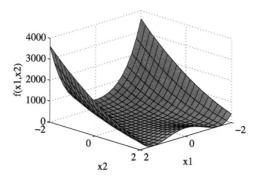

Figure A.12 Rosenbrock's function with two variables.

Rosenbrock's Function [ROS] This function, also known as De Jong's function 2, in n_{var} variables, is defined by:

$$f(\mathbf{x}) = \sum_{i=1}^{n_{var}-1} \left(100(x_{i+1} - x_i^2)^2 + (x_i^2 - 1)^2\right)$$

The global minimum of the function is at $\mathbf{x}^* = [1 \quad 1\ldots1]$ and $f(\mathbf{x}^*) = 0$. The function is usually searched in the hypercube $\mathbf{x} \in [-30, \ 30]$. A plot of this function in two variables is shown in Figure A.12.

Rosenbrock's Modified Function [ROSM] This function, in 2 variables, is defined by:

$$f(\mathbf{x}) = 74 + 100(x_2 - x_1^2)^2 + (1 - x_1)^2 - 400e^{-\frac{(x_1+1)^2+(x_2+1)^2}{0.1}}$$

This function, has a "bump" in the valley (compared with Rosenbrock's function). The function has a global minimum at $\mathbf{x}^* = [-0.9, \ -0.95]$, giving $f(\mathbf{x}^*) = 34.37$. This modification makes it a difficult to optimize because the local minimum basin is larger than the global minimum basin. The function is usually searched in the hypercube $\mathbf{x} \in [-2, \ 2]$. A plot of this function is shown in Figure A.13.

Spherical Function [SPH] This function, also known as De Jong's function 1, in n_{var} variables, is defined by:

$$f(\mathbf{x}) = \sum_{i=1}^{n_{var}} x_i^2$$

The global minimum of the function is at $\mathbf{x}^* = [0 \quad 0\ldots0]$ and $f(\mathbf{x}^*) = 0$. The function is usually searched in the hypercube $\mathbf{x} \in [-10, \ 10]$. A plot of this function is shown in Figure A.14.

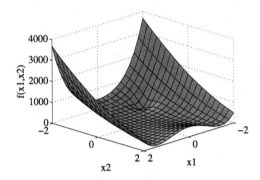

Figure A.13 Rosenbrock's modified function with two variables.

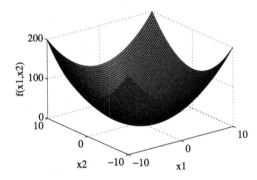

Figure A.14 Spherical function with two variables.

Styblinski–Tang Function   [STY] This function, in n_{var} variables, is defined by:

$$f(\mathbf{x}) = \frac{1}{2} \sum_{i=1}^{n_{var}} (x_i^4 - 16x_i^2 + 5x_i)$$

The global minimum of the function is at $\mathbf{x}^* = -2.903534[1, 1, \ldots 1]$ and $f(\mathbf{x}^*) = -39.16599n_{var}$. The function is usually searched in the hypercube $\mathbf{x} \in [-5, 5]$. A plot of this function in two variables is shown in Figure A.15.

Schwefel 2.26 Function   [SCH] This function, in n_{var} variables, is defined by:

$$f(\mathbf{x}) = 418.9829n_{var} - \sum_{i=1}^{n_{var}} x_i \sin(\sqrt{|\mathbf{x}|})$$

The global minimum of the function is at $\mathbf{x}^* = 420.9687[1, 1, \ldots 1]$ and $f(\mathbf{x}^*) = 0$. The function is usually searched in the hypercube $\mathbf{x} \in [-500, 500]$. A plot of this function in two variables is shown in Figure A.16.

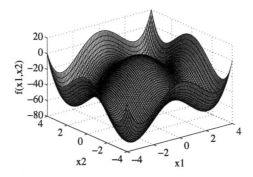

Figure A.15 Styblinski-Tang function with two variables.

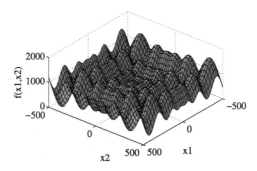

Figure A.16 Schwefel's function.

Program Listings

DiSCLAIMER.

While every effort has been made to provide functioning MATLAB scripts, the programs in this section are not guaranteed in any way or for any specific purpose. They are provided for use with the programming exercises given in this book and as such are written to provide clarity rather than efficiency. The use of these scripts requires some basic knowledge of MATLAB. Alternative versions are often available at the Mathworks web site for example.

This section provides programs for the ABC algorithm, the Cuckoo search algorithm and the Differential Evolution algorithm, together with required functions.

ABC algorithm. You must save this script in your MATLAB directory under an appropriate name.

```
% Matlab program for the ABC algorithm
% scripting guided by the paper by
% Karaboga & Basturk 2007
% J Glob Opt (2007) 39:459-471
%
% Scripteded in Matlab GL/JP October 2015
%
% Maximum number of generations maxgen
clear all
NR = input('Enter number of runs of the algorithm required ');
runaverage = 0;  runfset = [ ];  totgen = 0;
failrun = 0;
% ****************Key parametrs are set here******************
% Maximum number of generations
maxgen = 4000;
% Define function to be optimised
f = 'fg2test';
% Number of test function variables nv
nv = 10;
% Range of domain variables rmax to rmin
rmax = 5;  rmin = -5;
% Number of food sources
NS = 30;
% ***********************************************************
limit = nv*NS;
% Begin run loop
for runs = 1:NR
    gen = 1;
        fail = zeros(NS);
    beepop = rand(NS,nv);
```

Introduction to Nature-Inspired Optimization
DOI: 10.1016/B978-0-12-803636-5.00021-9

```
% Confine population variables within the domain of the problem
x = (rmax-rmin)*beepop+rmin;
xval0 = x;   xn = x;
% Calculate initial objective function values
fval = [ ];
fbestvals = [ ];
xlb = x;
flb = feval(f,xlb);
[fvorder,index] = sort(flb);
xgb = xlb(index(1),:);  fgb = feval(f,xgb);

fgbset=[ ];
% Start main loop
while(gen < maxgen)
    % Keep list of best values
    fbestvals = [fbestvals;fgb];
    % Assign employed bees
    for i = 1:NS
        xn(i,:) = x(i,:);
        % select random k
        % Randomly select k and make sure
        % k is not equal to i
        k = i;
        while k == i
            k = ceil(rand()*NS);
        end % while
        j = ceil(rand()*nv);
        xn(i,j) = x(i,j)+(2*rand()-1)*(x(i,j)-x(k,j));
        if xn(i,j) > rmax
            xn(i,j) = rmax;
        end
        if xn(i,j)<rmin
            xn(i,j)=rmin;
        end

        fn = feval(f,xn(i,:));
        xp = x(i,:);   fp = feval(f,xp);
        % calculate fitness
        if fp >= 0
            fitp = 1/(1+fp);
        else
            fitp = 1+abs(fp);
        end
        if fn >= 0
            fitn = 1/(1+fn);
        else
            fitn = 1+abs(fn);
        end
        if fitp < fitn
            x(i,:) = xn(i,:);
            fail(i) = 0;
        else
            fail(i) = fail(i)+1;
```

```
        end
end % NP
% Ensure new points are still within the problems
% region of definition
for i = 1:NS
    for k = 1:nv
        if x(i,k) > rmax
            x(i,k) = rmax;
        end
        if x(i,k) < rmin
            x(i,k) = rmin;
        end
    end % k
end % NP
% evaluate objective of all x values find best
fval = [ ];
fval = feval(f,x);
% Calculate selection probabilites for onlooker bees
% first calculate fitness function
for i = 1:NS
    if fval(i) >= 0
        fitf(i) = 1/(1+fval(i));
    else
        fitf(i) = 1+abs(fval(i));
    end
end %NP
% Now work out probabilities for onlooker bees
% Alternative method
% for i=1:NS
    % p(i)=fitf(i)/sum(fitf);
% end
% Alternative
fbest = max(fitf);
for i = 1:NS
    p(i) = 0.9*fitf(i)/fbest+0.1;
end
% Onlooker stage dependent on probability information select
% bee to change
for i = 1:NS
    if rand() < p(i)
        % select random k
        % Randomly select k and make sure
        % k is not equal to i
        k = i;
        while k == i
            k = ceil(rand()*NS);
        end % while
        j = ceil(rand()*nv);
        xn(i,j) = x(i,j)+(2*rand()-1)*(x(i,j)-x(k,j));
        if xn(i,j) > rmax
            xn(i,j) = rmax;
        end
        if xn(i,j) < rmin
```

```
                xn(i,j) = rmin;
            end
            fn = feval(f,xn(i,:));
            xp = x(i,:);   fp = feval(f,xp);
            % Compute fitness
            if fp >= 0
                fitp = 1/(1+fp);
            else
                fitp = 1+abs(fp);
            end
            if fn >= 0
                fitn = 1/(1+fn);
            else
                fitn = 1+abs(fn);
            end
            if fitp < fitn
                x(i,:) = xn(i,:);
                fail(i) = 0;
            else
                fail(i) = fail(i)+1;
            end
        end %if
    end %NP
% Scout bee phase scout bee searches solution
% space for new source location to replace
% unsatisfactory source
[maxfail,indexmax] = max(fail);
if maxfail >= limit
    x(indexmax,:) = (rmax-rmin)*rand(1,nv)+rmin;
    fail(indexmax) = 0;
    fval(indexmax) = feval(f,x(indexmax,:));
end
% Update function all values
fval = feval(f,x);
% Find best vaue and best point so far
fgb = min(flb);
for i = 1:NS
    if fval(i) < flb(i)
        xlb(i,:) = x(i,:);
        flb(i) = fval(i);
    end
    xlbv = xlb(i,:);
    flb(i) = feval(f,xlbv);
    if flb(i) < fgb
        xgb = xlbv;
        fgb = flb(i);
    end
end
fgb;
x = xlb;
flb = fval;
fgbset = [fgbset fgb];
gen = gen+1;
```

```
      end % generationloop
      if gen >= maxgen
          failrun = failrun+1;
      end
      disp('Final best solution at')
      disp('Iteration')
      disp(gen)
      disp('Optimum point ')
      xgb
      disp('Optimum Objective function value = ')
      disp(fgb)
      runfset = [runfset fgb];
      totgen = totgen+gen;
end % NR
% Calculate statistics for the runs:
maxfset = max(runfset);
minfset = min(runfset);
runaverage = mean(runfset);
stdev = std(runfset);
disp('Average of runs = ')
runaverage
disp('Best result = ')
minfset
disp('Worst result = ')
maxfset
disp('standard deviation = ')
stdev
```

Cuckoo Search algorithm.   You must save this script in your MATLAB directory under an appropriate name.

```
% Matlab program for the basic Cuckoo Search  Optimisation algorithm
% Yang, X. S.  and Deb, S. 2010. Cuckoo search via Levy Flights. in
% Proc. of World Congress on
% Nature & Biologically Inspired Computing, 2009. NaBIC 2009.
% December 2009, India. IEEE Publications, USA, pp. 210-214 (2009)
% Program icludes some aspects of the
% Includes modifications suggested by Tawfik et al. 2013
%
% Programmed by GL/JP 2015
%
% Maximum number of generations maxgen
clear all
NR = input('Enter number of runs of the algorithm required ');
runaverage = 0;  runfset = [ ];
totgen = 0;
failrun = 0;
% ****************Key parametrs are set here******************
% Maximum number of generations
maxgen = 4000;
% Define function to be optimised
f = 'fg4test';
% Set number of variables for function
```

```
nv = 10;
% Set initial number of nests
NP = 30;
% Set step size
s = 0.05;
% Range of domain variables rmax to rmin
rmax = 5; rmin = -5;
% ************************************************************
for runs = 1:NR
    gen = 1;
    % Set initial population matrix randomly
    cpop = rand(NP,nv);
    % Confine population variables within the
    % domain of the problem rmin to rmax
    x = (rmax-rmin)*cpop+rmin;   xval0 = x;
    % Set up initial values for global and local minimu
    xlb = x;
    flb = feval(f,xlb);
    [fvorder,index] = sort(flb);
    xgb = xlb(index(1),:);   fgb = feval(f,xgb);
    % Calculate initial objective function values
    fval = [ ];
    for i = 1:NP
        xp = x(i,:);
        fval(i) = feval(f,xp);
    end
    x0 = x;   oldfval = fval;
    fgbset = [ ];

    % Set constants PA is the proportion of nests abondoned, suggested
    % value 0.05 to 0.25, s is the step size factor,
    % These values can be tuned for individual types of problem
    % lambda is the
    % Levy search parameter suggested values 1 and 1.5.
    % alternative test set
    % PA = 0.05; s = 0.01;
    PA = 0.1;   s = 0.05;
    %**Start main loop**

    while(gen < maxgen)
        % Main steps in cuckoo search algorithm note
        % the Levy step is introduced using Levydist (1)
        % where Mantegna's algorithm has been used to
        % generate the Levy values.
        % Now update nestvalues
        for i = 1:NP
            for j = 1:nv
                x(i,j) = x(i,j)+s*Levydist(1)*randn()*(x(i,j)-xgb(j));
                % alternative
                % x(i,j) = x(i,j)+s*Levydist(1)*(x(i,j)-xgb(j));
            end % nvloop
        end % NPloop
        %Ensure points are confined to solution space
```

```matlab
for i = 1:NP
    for k = 1:nv
        if x(i,k) > rmax
            x(i,k) = rmax;
        end
        if x(i,k) < rmin
            x(i,k) = rmin;
        end
    end % k
end % NP
% Update function all values
fval = feval(f,x);fgb=min(flb);
for i = 1:NP
    if fval(i) < flb(i)
        xlb(i,:) = x(i,:);
        flb(i) = fval(i);
    end
    xlbv = xlb(i,:);
    flb(i) = feval(f,xlbv);
    if flb(i) < fgb
        xgb = xlbv;
    end
end
x = xlb;
%Keep set of best values so far
fgbset = [fgbset fgb];
%Replace fraction of nests pa
RP1 = randperm(NP); RP2 = randperm(NP);
for i = 1:NP
    for j = 1:nv
        if rand() <= PA
            x(i,j) = x(i,j)+rand()*(x(RP1(i),j)-x(RP2(i),j));
        end % if
    end % nv
end % NP
% Ensure solution is in solution space
for i = 1:NP
    for k = 1:nv
        if x(i,k) > rmax
            x(i,k) = rmax;
        end
        if x(i,k) < rmin
            x(i,k) = rmin;
        end
    end % k
end % NP
% Update objective function values
fval = feval(f,x);   fgb = min(flb);
for i = 1:NP
    if fval(i) < flb(i)
        xlb(i,:) = x(i,:);
        flb(i) = fval(i);
    end
```

```
                xlbv = xlb(i,:);
                flb(i) = feval(f,xlbv);
                if flb(i) < fgb
                    xgb = xlbv;
                    fgb = flb(i);
                end
            end
        x = xlb;
        gen = gen+1;
    end %generationloop
    if gen >= maxgen
        failrun = failrun+1;
    end
    disp('best approximation to minimum point')
    xgb
    disp('best approximation to minimum value')
    fgb
    disp('Number of generations')
    gen
    runfset = [runfset fgb];
    totgen = totgen+gen;
end % NR
maxfset = max(runfset);
minfset = min(runfset);
runaverage = mean(runfset);
stdev = std(runfset);
disp('Average of runs = ')
runaverage
disp('Best result = ')
minfset
disp('Worst result = ')
maxfset
disp('standard deviation = ')
stdev
gen
avgen = totgen/NR
```

Levydist Function. This function must be saved as Levydist in your MATLAB directory. It is called by the cuckoo search script.

```
function res = Levydist(c)
% Generating samples from the Levty distribution using,
% the Mantegna algorithm, reference
% R.N. Mantegna, 'Fast, accurate algorithm for numerical simulation of
% Levy stable stochastic processes,'Physical Review vol 49 no 4
% pp 4677-4683,1994
% The parameter c governs the shape of the curve,c>0
% this function returns a single value
% but is easily modified to produce a vector of values.
% Suggested values of c  are 1 and 1.5. GRL/JP
c1 = 1+c;   rc = 1/c;
sigmau = gamma(c1)*sin(pi*c/2)/(gamma(c1/2)*c*2^((c-1)/2));
sigmau = sigmau^rc;
```

```
u = sqrt(sigmau)*randn;
v = randn;
res = u./abs(v).^rc;
```

The following MATLAB functions must be saved in tour MATLAB directory under the names given, i.e. fg12test, fg2test and fg4test. These functions implement the Egg crate function, the Rosenbrock function and the Rastrigin function respectively.

Function fg12test.   Eggcrate function

```
function objval = fg12test(x)
% Eggcrate function test function
% Solution xi = 0 all i; and global optimum = 0
% Each variable range range -5 to 5
% See APPENDIX A for a description of this function
% NB this will calculate a set function values
[Nset, Nvar]  =  size(x);
S = 0;
for i = 1:Nvar
    S = S+x(1:Nset,i).^2+25*sin(x(1:Nset,i)).^2;
end
objval = S;
objval = objval';
```

Function fg2test.   Rosenbrock's function

```
function objval = fg2test(x)
% Rosenbrocks test function
% Solution x = (1,1) and objval = 0
% Number of variables 2 range -5 to 5
% See APPENDIX A for a description of this function
% NB this will calculate a set function values
[Nset, nv]  =  size(x);
objval = 0;
for i = 1:nv-1
    objval = objval+(1-x(1:Nset,i)).^2+100*(x(1:Nset,i+1)
            -x(1:Nset,i).^2).^2;
end
objval = objval';
```

Function fg4test.   Rastrigin's function

```
function objval = fg4test(x)
% Rastrigin test function
% Solution xi = 0 all i; and global optimum = 0
% Number of variables 10 range -5 to 5
% See APPENDIX A for a description of thisfunction
% NB this will calculate a set of function values
[Nset, Nvar]  =  size(x);
S = 0;
for i = 1:Nvar
```

```
    S = S+x(1:Nset,i).^2-10*cos(2*pi*x(1:Nset,i));
end
objval = 10*Nvar+S;
objval = objval';
```

Differential evolution. You must save this script in your MATLAB directory under an appropriate name.

```
% Differential Evolution DE/rand/1/bin
% Equation numbers refer to Storn & Price Paper
%
% Scripteded in Matlab GL/JP October 2015
%
% x0_min is a column vector specifying the minimum value of each variable
% in the search space
% x0_max is a column vector specifying the maximum value of each variable
% in the search space
% F is user defined mutation parameter 0.85 suggested
% C is user defined crossover parameter 0.5 suggested
% nvar is the numer of variables or dimensions
% npop is the size of the population
% ngen is the maximum number of generations for the search
clear all

%%%%%%%%%%%%%%%%%%%%%%%%%%%%%%%%%%%%%%%%%%%%%%%%%%%%%%%%%%%%%%%%%%%%%%
% Need to insert function to be minimised, with necessary parameters. %%
funk= @(x) ROSDE(x)                                                  %%
nvar = 10;                                                           %%
x0_min = -5*ones(nvar,1);                                            %%
x0_max =  5*ones(nvar,1);                                            %%
%o%pt_f = 0;                                                         %%
F = 0.85;                                                            %%
C = 0.5;                                                             %%
npop = 30;                                                           %%
ngen = 4000;                                                         %%
nrun = 10;                                                           %%
%%%%%%%%%%%%%%%%%%%%%%%%%%%%%%%%%%%%%%%%%%%%%%%%%%%%%%%%%%%%%%%%%%%%%%
%-----------------------------------------------------------------
x_rng = x0_max-x0_min;

for nr = 1:nrun
    % Generate initial locations for population
    for i = 1:npop
        x(1:nvar,i) = x0_min+x_rng.*rand(nvar,1);
        x_max(:,i) = x0_max;
        x_min(:,i) = x0_min;
    end

    for gen = 1:ngen
        % Mutation
        for i = 1:npop
            flg = 0;
            while flg == 0
```

```
            r = randi(npop);
            if r~=i
                r1 = r; flg = 1;
            end
        end

        flg = 0;
        while flg == 0
            r = randi(npop);
            if r~=i & r~=r1
                r2 = r;   flg = 1;
            end
        end

        flg = 0;
        while flg == 0
            r = randi(npop);
            if r~=i & r~=r1 & r~=r2
                r3 = r; flg = 1;
            end
        end

        v(:,i) = x(:,r1)+F*(x(:,r2)-x(:,r3));  % Equation (2)
end   % End populations loop

r = rand(nvar,npop);
ri = randi(nvar,npop,1);

% Crossover and selection
for j = 1:npop
    % Mutation
    for i = 1:nvar
        % Line below is compact for of Equation (4)
        u(i,j) = v(i,j)*(r(i,j)<=C | j==ri(i,1))+x(i,j)*(r(i,j)>C
                & j~=ri(i,1));
    end
    % Selection
    if funk(u(:,j))<funk(x(:,j))
        x(1:nvar,j) = u(1:nvar,j);
    end
end % End populations loop

% Limit x to search area.
for i = 1:nvar
    for j = 1:npop
        if x(i,j) > x_max(i)
            x(i,j) = x_max(i);
        end
        if x(i,j) < x_min(i)
            x(i,j) = x_min(i);
        end
    end
end
```

```
    end  % End generations loop

    for i = 1:npop
        res(i) = funk(x(:,i));
    end

    fprintf('run number = %4.0f\n', nr)
    [min_r, idx] = min(res);
    min_res(nr) = min_r;
    min_x(:,nr) = x(:,idx);
    sol_x = min_x(:,nr)'
    fprintf('min_fun = %16.8e\n', min_res(nr))

    disp('-----------------------------------')
end % End runs loop

[min_min_res,kdx] = min(min_res);

p = min_x(:,kdx);

fprintf(' actual soln = %18.8e\n', opt_f)
fprintf(' min_min_res = %18.8e\n', min(min_res))
fprintf(' max_min_res = %18.8e\n', max(min_res))
fprintf('mean_min_res = %18.8e\n', mean(min_res))
fprintf(' std_min_res = %18.8e\n', std(min_res))
fprintf('F_value = %4.2f, c_value = %4.2f, Gens = %4.0f,', F, C, ngen)
fprintf('  Pop = %3.0f, Run = %3.0f, Dimensions = %3.0f\n', npop,
    nrun, nvar)
fprintf('Solution for minimum value of function\n')
for i = 1:nvar
    fprintf('%18.8e\n',p(i))
end
```

The following MATLAB functions must be saved in your MATLAB directory under the names given, i.e. RASDE, ROSDE and EGGDE. These functions implement the Rastrigin function, the Rosenbrock function and the Egg crate function respectively.

Function RASDE. `function val = RASDE(x)`
```
%Rastrigin function for nv varables
[nv,p] = size(x);
sum = 0;
for i = 1:nv
    sum = sum+10+x(i).^2-10*cos(2*pi*x(i));
end
val = sum;
end
```

Function ROSDE. `function val = ROSDE(x)`
```
%Rosenbrocks function for nv varables
[nv, p] = size(x);
sum = 0;
for i = 1:nv-1
```

```
        sum = sum+(x(i)-1).^2+100*(x(i+1)-x(i).^2).^2;
    end
    val = sum;
    end
```

Function EGGDE.
```
                        function val = EGGDE(x)
    %Egg crate function for nv varables
    [nv,p] = size(x);
    sum = 0;
    for i = 1:nv
        sum = sum++x(i).^2+25*sin(x(i)).^2;
    end
    val = sum;
    end
```

SOLUTIONS TO PROBLEMS

CHAPTER 1

1.1. When $x_1 = x_2 = x_3 = 1$, the gradient vectors are $[2\,2\,2]$, $[-2\sin(4)\ -2\sin(4)]$ and $[-2/3\ -2/3]$ for the functions f_1, f_2 and f_3 respectively.

1.3. $\lambda = 36/136 = 0.265$ and $f(\mathbf{x}^{(1)}) = 0.3102$.

1.4. $\lambda = 148/872 = 0.17$ and $\mathbf{x}^{(1)} = [0.34\ 2.04]$.
$\lambda = 180/383 = 0.47$ and $\mathbf{x}^{(2)} = [0.96\ 1.93]$.

1.6. When $m = 6$ the sequence of pseudo-random numbers generated is 0, 5, 4, 3, 2, 1. The sequence then repeats. When $m = 9$ the sequence of pseudo-random numbers generated is 0, 2, 7, 6, 8, 4, 3, 5, 10. The sequence then repeats.
Note that in general the values repeat themselves after the mth value.

1.7. (i) $10\pi \cos(10\pi x)/(2x) - \sin(10\pi x)/(2x^2) = 0$
(ii) $(6x - 2)\sin(12x - 4) + (6x - 2)^2 \cos(12x - 4) = 0$

1.8. The minimum of the function is $f = -72$ when $x = 3$, $y = -1$.

CHAPTER 2

2.1. For decimal 10 to 20,

decimal	binary	Gray	decimal	binary	Gray
10	01010	01111	16	01000	11000
11	01011	01110	17	10001	11001
12	01100	01010	18	10010	11011
13	01101	01011	19	10011	11010
14	01110	01001	20	10100	11110
15	01111	01000			

2.2.

x	$f(x)$	x
5	15	101
6	16	110
4	20	100
7	23	111
3	31	011
2	48	010
1	71	001

2.3. The fittest pair are

$$\begin{array}{c|cc} x & 4.9 & 0.7 \\ y & 1.4 & 3.7 \\ f(x, y) & 811 & 696 \end{array}$$

(1) By crossover

$$\begin{array}{c|cc} x & 4.9 & 0.7 \\ y & 3.7 & 1.4 \\ f(x, y) & 547 & 703 \end{array}$$

(2) Using (2.2) with $r = 0.3$,

$$\begin{array}{c|cc} x & 1.96 & 3.64 \\ y & 3.01 & 2.09 \\ f(x, y) & 719 & 608 \end{array}$$

(3) Using (2.3), (2.4) and (2.5).

$$\begin{array}{c|ccc} x & 2.80 & 7.00 & -1.40 \\ y & 2.55 & 0.25 & 4.85 \\ f(x, y) & 712 & 782 & 699 \end{array}$$

Note that the third child is infeasible and must be rejected.

2.4.

x	8.1000	1.3000	6.3000	2.8000	9.6000	1.6000
f_x	2.7604	1.0986	0.0422	0.5605	−0.5401	1.2644
v	9.3000	1.9000	0.1000	3.1000	4.6000	5.1000
u	9.3000	1.9000	0.1000	2.8000	9.6000	5.1000
f_u	0.3795	1.3044	0.0316	0.5605	−0.5401	−2.0908
x_n	9.3000	1.3000	0.1000	2.8000	9.6000	5.1000

CHAPTER 3

3.1. The minimum value is at $f(\mathbf{x}_2) = 5$, therefore $x_{gopt} = [2, 1]$, but at first iteration this is the same as x_{lopt}.

3.2. $v_1^{(t+1)} = (-0.325, 0)$ and $v_2^{(t+1)} = (0, 0)$. $x_1^2 = (2.5, 1) + (-0.325, 0) = (2.175, 1)$ and $x_2^2 = (2, 1)$. On calculating the new objective function values there is no change in the current global and local optimum values.

3.3. $w^{(1)} = 0.92$, $w^{(2)} = 0.84$, $w^{(3)} = 0.76$, $w^{(4)} = 0.68$, $w^{(5)} = 0.60$ ..., $w^{(10)} = 0.20$.

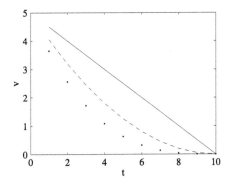

Figure 1 Problem 3.4. Plot for $p = 1$ (full), $p = 2$ (dash) and $p = 3$, (dots).

3.4. The three graphs of (3.11) for values $p = 1$, 2, 3. $p = 1$ gives a linear decrease $p = 2$, 3 give nonlinear decrease. See Figure 1.

CHAPTER 4

4.1. We obtain the new nest locations at (1.015 0.95), (−0.875 2.025), (0.15 1.2), (1.45 2.05). The function values are $f(\mathbf{x}_1^1) = 1.935$, $f(\mathbf{x}_2^1) = 4.867$, $f(\mathbf{x}_3^1) = 1.463$, and $f(\mathbf{x}_4^1) = 6.305$. The minimum value of these is $f(\mathbf{x}_3^1) = 1.463$. Ranking them we have $f(\mathbf{x}_3^1), f(\mathbf{x}_1^1), f(\mathbf{x}_2^1)$ and the worst is $f(\mathbf{x}_4^1)$. On comparison with the objective function of the original values we obtain a new set of locations, (1.015 0.95), (−0.875 2.025), (0 1) and (1.5 2).

4.3. The gradients are: (i) $(2x_1, 2x_2)$ (ii) $(-\sin(x_1 + x_2), -\sin(x_1 + x_2))$ (iii) $(-x_2 \sin(x_1 x_2), x_1 \sin(x_1 x_2))$ Hence the values of the gradient at the given points are $(1, 1)$, $(-sin(1), -\sin(1))$ and $(0.5 \sin(0.25), 0.5 \sin(0.25))$.

4.4. $P(1) = 1.8$, $P(2) = 1.6$, $P(3) = 1.4, ..., P(10) = 0$.

4.5. First calculate c, $c = 20 \log_e(0.1/1)$ since $t_{max} = 20$, $c = -2.3025$. Thus $\alpha_t = 1 \times \exp(-2.3025t)$ for $t = 1, 2, ..., 20$ (see Figure 2).

CHAPTER 5

5.1. See Figure 3.

5.2. See Figure 4.

5.3. Partial solution: distance between firefly 1 and others $\sqrt{2}$, 1, $\sqrt{5}$ and $2\sqrt{2}$. Average, $d_1 = (3\sqrt{2} + 1 + \sqrt{5})/4$ distance between firefly 2 and others $d_2 = (2\sqrt{2} + 2)/4$

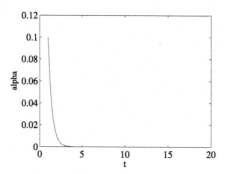

Figure 2 Solution for Problem 4.5.

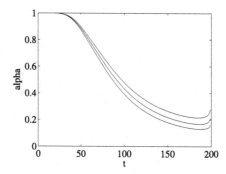

Figure 3 Solution for Problem 5.1.

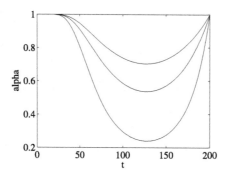

Figure 4 Solution for Problem 5.2.

calculate d_3 and d_4 objective function values at each point are $f(1, 1) = 2, f(2, 2) = 8,$ $f(1, 2) = 5, f(3, 2) = 13, f(3, 3) = 18,$ minimum at point $(1, 1)$. Can calculate d_{opt} hence D.

5.4. Using the logistic map with $r = 2$ we have, $x_0 = 0.25$, $x_1 = 0.375$, $x_2 = 0.46875$, $x_3 = 0.49805$, $x_4 = 0.49999$, $x_5 = 0.49999$. When $r = 4$, $x_0 = 0.1$ we have $x_1 =$

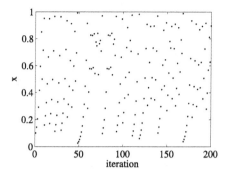

Figure 5 Solution for Problem 5.5.

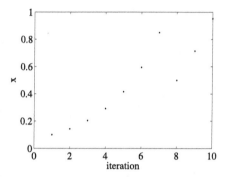

Figure 6 Solution for Problem 5.6.

0.36, $x_2 = 0.9216$, $x_3 = 0.2890$, $x_4 = 0.8219$, $x_5 = 0.5855$. Values generated using formula: i.e. 0.3600, 0.9216, 0.2890, 0.8219, 0.5854, and 0.3750, 0.4688, 0.4980, 0.5000, 0.5000, show good agreement.

5.5. The program you have written should generate a figure similar to Figure 5.

5.6. Using the asymmetric tent map we have for the values given $m = 0.7$: $x_1 = 0.1429$, $x_2 = 0.2041$, $x_3 = 0.2916$, $x_4 = 0.4166$, $x_5 = 0.5951$, $x_6 = 0.8501$. Note how formula changes at this stage since $m < x_6$. $x_7 = 0.4997$. The formula changes again as $x_7 < m$ $x_8 = 0.7139$. Changes back $x_9 = 0.9537$, $x_{10} = 0.1533$. Now plotting these 10 values gives for $m = 0.7$ and produces Figure 6.

CHAPTER 6

6.1. The current bacteria we are considering is at [1.5, 2.5] this is interacting with the other bacteria [1, 1], [2, −1], [2, 0] and [2, 3] gives these values are 2.5, 12.5, 6.5,

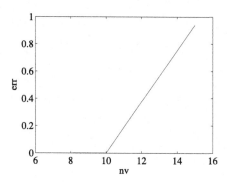

Figure 7 Solution for Problem 7.1.

0.5. Thus the first term becomes: $-0.1 \times [\exp(-0.2 \times 2.5) + \exp(-0.2 \times 12.5) + \exp(-0.2 \times 6.5) + \exp(-0.2 \times 0.5)]$ and the second term $0.1 \times [\exp(-3 \times 2.5) + \exp(-3 \times 12.5) + \exp(-3 \times 6.5) + \exp(-3 \times 0.5)$. Sum gives J_{α} objective function is $f(1.5, 2.5) = 2(1.5)^2 + (0.5)^2 = 4.75, J = 4.75$. From (6.3) we have $J_{ar} = \exp(M - 4.75)J_{\alpha}$.

6.2. Setting the random vector arbitrarily as $u = [0.5, 0.25]$, $x_0 = [1, 1.5]$, $x_1 = [1.0895, 1.5447]$. Repeating this again with new arbitrary **u** vector: (In practice this repetition only occurs if improvement is achieved.) $x_2 = [1.0895, 1.5447] + [0.1, 0.1] \times [0.75, 0.50]/0.9014$. Thus $x_2 = [1.9215, 2.08954]$.

CHAPTER 7

7.1. The errors in each case are given by:

$$[0, 1.6045 \times 10^{-8}, 9.3491 \times 10^{-4}, 0.9401]$$

See Figure 7.

7.2. Taking, $j = 1$, $c_{0,1} = 0.10$, $\mu = 0.5$, and $K = 3$ we have: for $k = 0$ we have $c_{1,1} = 0.045$, for $k = 1$ we have $c_{2,1} = 0.0215$, for $k = 2$ we have $c_{3,1} = 0.0105$, for $k = 3$ we have $c_{4,1} = 0.0052$.

Now repeating this for $j = 2$ we have for $k = 0$ we have $c_{1,2} = 0.0938$, for $k = 1$ we have $c_{2,2} = 0.0425$, for $k = 2$ we have $c_{3,2} = 0.0203$, for $k = 3$ we have $c_{4,2} = 0.0099$.

7.3. We begin by calculating the objective function value for bees (2, 5), (3, 2), (1, 1) and (6, 2) labeled as points 1, 2, 3, 4 so that for point 1 $f(2, 5) = 4.996$ for point

2 $f(3,2) = 9.999$ for point 3 $f(1,1) = 1.999$ for point 4 $f(6,2) = 36.999$. Calculate fitness, since in each of the above cases $f(x) > 0$ we have $fit_1 = 0.1667$, $fit_2 = 0.0909$, $fit_3 = 0.3333$ and $fit_4 = 0.0263$. Sum of the fitnesses is 0.6172. Therefore: $p_1 = 0.27$, $p_2 = 0.1473$, $p_3 = 0.5400$ and $p_4 = 0.0426$. Summing these we obtain 0.9999 as expected approximately unity.

7.4. The traveling salesman problem we first designate the variable x_{ij} to represent if the part of the route from node i to j is taken or not. Thus $x_{ij} = 1$ part of the route is taken but $x_{ij} = 0$ if it is not. Now the aim is to minimize the length of the route but this can be expressed in terms of x_{ij} and L_{ij} as follows:

$$\text{Minimize} \sum_{i=0}^{n} \sum_{j=1, j \neq i}^{n} L_{ij} x_{ij}$$

That is the sum of all the route segments. However this is subject to the constraints that all towns must be visited these are that each city or town must be arrived at from and departed to only one city or town so the constraints are represented by

$$\sum_{i=0, i \neq j}^{n} x_{ij} = 1$$

and

$$\sum_{j=0, j \neq i}^{n} x_{ij} = 1$$

also $x_{ij} = 0$ or 1.

7.5. Using the given formula we have $p_1 = 0.4566$. Now since $p_1 < 0.5$, this ant takes the other route $p_2 = 0.4364$, $p_3 = 0.4172$ and $p_4 = 0.3990$.

CHAPTER 8

8.1. $T = 1$, $x_0 = 0.25$, $f(x_0) = 0.1237$. Iteration 1. Select $x_n = 1.25$, $f(x_n) = 1.0610$, $f(x_n) - f(x_0) > 0$. Test for acceptance, $\exp((0.1237 - 1.0610)/1) = 0.3917 > 0.32$. Therefore $x_0 = x_n = 1.25$. Iteration 2. New value $x_n = 2.5$, $f(x_n) = 0.9463$. $f(x_n) - f(x_0) < 0$, therefore $x_0 = x_n = 2.5$. $T = 0.1$. Iteration 1. Select new random value $x_n = 2.7$, $f(x_n) = 0.7032$. $f(x_n) - f(x_0) < 0$, therefore accept $x_0 = x_n = 2.7$. Iteration 2. Select new random point $x_n = 3$, $f(x_n) = 0.2444$. $f(x_n) - f(x_0) < 0$. Therefore accept $x_0 = x_n = 3$.

8.2. Iteration 1.

x	8.1	1.3	6.3	2.8	9.6	1.6
y	9.1	9.1	1.0	5.5	9.6	9.7
f	3.7230	2.0612	0.8837	-1.0941	-1.0803	0.4180

$f_{mean} = 0.8186$, $x_c = 0.9506$, $y_c = 7.0890$.
Iteration 2.

x	2.7756	2.7256	0.6506	4.4756	2.6256	2.7506
y	6.9390	6.5640	5.0000	5.0000	4.0640	5.0000
f	2.2025	1.3770	-1.6557	-4.2008	-1.0803	-1.5122

$f_{mean} = -0.8116$, $x_c = 7.1465$, $y_c = 6.7954$.

8.3. The positions of the agents after one iteration are [7.4691, 8.4868], [1.6229, 8.4052], [5.9042, 1.5434] and [2.9013, 5.4118]. These positions are close but not identical to the initial positions.

8.4. The positions of the probes after one iteration are [2.1712, 3.9760], [3.9011, 6.6521], [6.3734, 4.5200] and [7.0000, 8.0000]. Note that the fittest and therefore most massive probe does not move.

REFERENCES

Abdelaziz, A., Mekhamer, S., Badr, M., Algabalawy, M., 2015. The firefly meta-heuristic algorithm developments and applications. International Electrical Engineering Journal 6 (7), 1945–1952.

Abraham, S., Sanyal, S., Sanglikar, M., 2013. Finding numerical solutions of Diophantine equations using ant colony optimization. Mathematics and Computation Archive 219 (24), 11376–11387.

Adby, P., Dempster, M., 1974. Introduction to Optimization Problems. Chapman and Hall, London.

Adoria, P., Diliman, U., 2005. MVF—multivariate test functions library in C for unconstrained global optimization list 54 functions. Retrieved January 23, 2017, from http://www.geocities.ws/eadorio/mvf.pdf.

Alfi, A., Modares, H., 2011. System identification and control using particle swarm optimization. Applied Mathematical Modelling 35 (3), 1210–1221.

Ali, M., Pant, M., Nagar, A., 2011. Two new approach incorporating centroid based mutation operators for differential evolution. World Journal of Modelling and Simulation 7 (1), 16–28.

Arasomwan, M., Adewumi, A., 2013. On the performance of linear decreasing inertia weights particle swarm optimization for global optimization. The Scientific World Journal, 860289.

Asi, M., Dib, N., 2010. Design of multilayer microwave broadband absorbers using central force optimization. Progress In Electromagnetics Research B 26, 101–113.

Babua, B., Munawarb, S., 2007. Differential evolution strategies for optimal design of shell-and-tube heat exchangers. Chemical Engineering Science 62 (14), 3720–3739.

Bacanin, N., Pelevic, B., Tuba, M., 2014. Constrained portfolio selection using artificial bee colony (ABC) algorithm. International Journal of Journal of Mathematical Models and Methods in Applied Science 8, 190–198.

Van den Berg, F., Engelbrecht, A., 2002. A new locally convergent particle swarm optimizer. In: Proceeding of the IEEE International Conference on Systems, Man and Cybernetics, pp. 96–101.

Van den Berg, F., 2006. Analysis of Particle Swarm Optimization. PhD Thesis. University of Pretoria, South Africa.

Biswas, A., Dasgupta, S., Das, S., Abraham, A., 2007. Synergy of differential evolution and bacterial foraging for global optimization. Neural Network World: International Journal on Neural and Mass-Parallel Computing and Information Systems 17 (6), 607–626.

Biswas, A., Mishra, K., Tiwari, S., Misra, A., 2013. Physics-inspired optimization algorithms: a survey. Journal of Optimization 2013, 438152.

Bolaji, A., Al-Betar, M., Awadallah, M., Khader, A., Abualigah, L., 2016. A comprehensive review: krill herd algorithm (KH) and its applications. Applied Soft Computing 49, 437–446.

Bolaji, A., Khader, A., Al-Betar, M., Awadallah, L., 2013. Artificial bee colony algorithm, its variants and applications. A survey. Journal of Theoretical and Applied Information Technology 47 (2), 434–459.

Brünger, A., Adams, P., Rice, L., 1997. New applications of simulated annealing in X-ray crystallography and solution NMR. Structure 5 (3), 325–336.

Camp, C., Pezeshk, S., Cao, G., 1998. Optimized design of two-dimensional structures using genetic algorithms. Journal of Structural Engineering 124 (5), 551–559.

Caranic, B., Fryer, C., Baines, R., 2001. An application of simulated annealing to the optimum design of reinforced concrete retaining structures. Computers and Structures 79 (17), 1569–1581.

de Castro, L., 2007. Fundamentals of natural computing: an overview. Physics of Life Reviews 4 (1), 1–36.

Černeý, V., 1985. A thermodynamic approach to the travelling salesman problem: an efficient simulation. Journal of Optimization Theory and Applications 45 (1), 41–51.

Chatterje, A., Siarry, P., 2006. Nonlinear inertia weight variation for dynamic adaption in particle swarm optimization. Computers and Operations Research 33 (3), 859–871.

Chauhan, N., Ravi, V., Chandra, D., 2009. Differential evolution trained wavelet neural networks: application to bankruptcy prediction in banks. Expert Systems with Applications 36 (4), 7659–7665.

Chen, W., 2014. An artificial bee colony algorithm for uncertain portfolio selection. The Scientific World Journal 2014, 578182.

Chen, Y., Yu, J., Mei, Y., Wang, Y., Su, X., 2016. Modified central force optimization (MCFO) algorithm for 3D UAV path planning. Neurocomputing 171 (1), 878–888.

Cheung, N., Ding, X.M., Shen, H.B., 2014. Adaptive firefly algorithm: parameter analysis and its application. PLOS ONE 9 (11), e112634.

Clerc, M., Kennedy, J., 2002. The particle swarm—explosion stability and convergence in a mullti-dimension complex space. IEEE Transactions on Evolutionary Computation 6 (1), 58–73.

Coley, D., 1999. An Introduction to Genetic Algorithms for Scientists and Engineers. World Scientific, Singapore.

Cuevas, E., Reyna-Orta, A., 2014. A cuckoo search algorithm for multimodal optimization. The Scientific World Journal 2014, 497514.

Daniel, W., 1990. Applied Nonparametric Statistics, 2nd ed.. PWS-Kent, Boston, pp. 262–274.

Das, S., Suganthan, P., 2011. Differential evolution: a survey of the state-of-the-art. IEEE Transactions on Evolutionary Computation 15 (1), 4–31.

Dasgupta, S., Das, S., Abraham, A., Biswas, A., 2009. Adaptive computational chemotaxis in bacterial foraging optimization: an analysis. IEEE Transactions on Evolutionary Computation 13 (4), 919–941.

Dash, M., Mohenty, R., 2014. Cuckoo search algorithm for speech recognition. International Journal of Advanced Research in Computer Engineering and Technology 3 (10), 3540–3545.

Davidon, W., 1959. Variable Metric Method for Minimisation. Report ANL 5990 (rev). Argonne Nation Laboratory, Illinois, USA.

Deb, K., 2000. An efficient constraint handling method for genetic algorithms. Computer Methods in Applied Mechanics and Engineering 186 (2–4), 311–338.

Deb, K., 2001. Multi-Objective Optimization Using Evolutionary Algorithms. John Wiley & Sons Inc., New York, USA.

Dorigo, M., 2007. Ant colony optimization. Scholarpedia 2 (3), 1461.

Dorigo, M., Gambardella, L., 1997. Ant colonies for the travelling salesman problem. Biosystems 43 (2), 73–81.

Dorigo, M., Maiezzo, V., Colorni, A., 1996. The ant system: optimization by a colony of cooperating agents. IEEE Transactions on Systems, Man, and Cybernetics, Part B (Cybernetics) 26 (1), 29–41.

Eberhart, R., Shi, Y., 2000. Comparing inertia weights and constriction factors in particle swarm optimization. In: Proceedings of the 2000 Congress on Evolutionary Computation. Evolutionary Computation.

Engelbrecht, A., 2005. Particle Swarm Optimization: Pitfalls and Convergence Aspects. PSO tutorial. Department of Computer Science, University of Pretoria.

Erol, O., Eksin, I., 2006. A new optimization method: Big Bang-Big Crunch. Advances in Engineering Software 37 (2), 106–111.

Fan, H., 2002. A modification to particle swarm optimization algorithm. Engineering Computations 19 (8), 970–989.

Fateen, S.K., Bonilla-Petriciolet, A., 2014. Gradient-based cuckoo search for global search. Mathematical Problems in Engineering 2014, 493740.

Feng, Y., Teng, G.F., Wang, A.X., Yao, Y.M., 2007. Chaotic inertia weight particle swarm optimization. In: Proceedings of the 2nd International Conference on Innovative Computing, Information and Control. Kumamoto, Japan.

Fiacco, A., McCormick, G., 1968. Nonlinear Programming Sequential Unconstrained Minimization Techniques. Wiley, New York.

Fiacco, A., McCormick, G., 1990. Nonlinear Programming Sequential Unconstrained Minimization Techniques. SIAM Classics in Mathematics. SIAM, Philadelphia (reissue).

Fister Jr., I., Yang, X.S., Fister, I., Brest, J., Fister, D., 2013. A brief review of nature-inspired algorithms for optimization. Elektrotehniski Vestnik 80 (3), 116–122.

Fister, I., Yang, X.-S., Brest, J., Fister Jr., I., 2014. On the randomised firefly algorithm. In: Yang, Xin-She (Ed.), Cuckoo Search and Firefly Algorithm Theory and Applications. In: Studies in Computational Intelligence, vol. 516, pp. 27–48.

Fletcher, R., Powell, M., 1963. A rapid convergent descent method for minimization. The Computer Journal 6 (2), 163–180.

Fletcher, F., Reeves, R., 1964. Function minimization by conjugate gradients. The Computer Journal 7 (2), 149–160.

Fogel, D., Ghozeil, A., 1997. Schema processing under proportional selection in the presence of random effects. IEEE Transactions on Evolutionary Computation 1 (4), 290–293.

Formato, R., 2007. Central force optimization: a new metaheuristic with applications in applied electromagnetics. Progress in Electromagnetics Research 77, 425–491.

Formato, R., 2010. Improved CFO algorithm for antenna optimization. Progress In Electromagnetics Research B 19, 405–425.

Formato, R., 2013. Pseudo-randomness in central force optimization. British Journal of Mathematics and Computer Science 3 (3), 241–264.

Forrester, A., Sobester, A., Keane, A., 2008. Engineering Design Via Surrogate Modelling: A Practical Guide. Wiley.

Friedman, M., 1937. The use of ranks to avoid the assumption of normality implicit in the analysis of variance. Journal of the American Statistical Association 32 (200), 675–701.

Friedman, M., 1940. A comparison of alternative tests of significance for the problem of m rankings. The Annals of Mathematical Statistics 11 (1), 86–92.

Friswell, M., Penny, J., Garvey, S., 1998. A combined genetic and eigensensitivity algorithm for the location of damage in structures. International Journal of Computers and Structures 69 (5), 547–556.

Gandomi, A., Alavi, A., 2012. Krill herd: a new bio-inspired optimization algorithm. Communications in Nonlinear Science and Numerical Simulation 17 (12), 4831–4845.

Gao, Y., Duan, Y., 2007. A new particle swarm algorithm with inertia weight and evolution strategy. In: Proceedings of the International Conference on Computational Intelligence and Security Workshops. Heilongjiang, China, pp. 199–203.

Gao, W., Liu, S., 2011. Improved artificial bee colony algorithm for global optimization. Information Processing Letters 111 (17), 871–882.

Gauci, M., Dodd, T., Grobe, R., 2012. Why 'GSA: a gravitational search algorithm' is not genuinely based on the law of gravity. Natural Computing 11 (4), 719–720.

Geem, Z., 2010. Research commentary: survival of the fittest algorithm or the novelest algorithm? International Journal of Applied Metaheuristic Computing 1 (4), 75–79.

Ghalia, M., 2008. Particle swarm optimization with an improved exploration-exploitation balance. In: IEEE Circuits and Systems. 51st Midwest Symposium on Circuits and Systems, pp. 759–762.

Gökdağ, K., Yildiz, A., 2012. Structural damage detection using modal parameters and particle swarm optimization. Materials Testing 54 (6), 416–420.

Goldberg, D., 1989. Genetic Algorithms in Search, Optimization and Machine Learning. Addison-Wesley Longman Publishing Co., Inc., Boston, MA, USA.

Goss, S., Aron, S., Deneubourg, J., Pasteels, J., 1989. Self organized shortcuts in the Argentine ant. Naturwissenscaften 76 (12), 579–581.

Gramacy, R., Lee, H., 2012. Cases for the nugget in modeling computer experiments. Statistics and Computing 22 (3), 713–722.

Green, J., Whalley, J., Johnson, C., 2004. Automatic programming with ant colony optimization. In: Proceedings of the 2004 UK Workshop on Computational Intelligence. Loughborough University, UK.

Gu, B., Pan, F., 2013. Modified gravitational search algorithm with particle memory ability and its application. International Journal of Innovative Computing, Information and Control 9 (11), 4531–4544.

Haghighi, A., Ramos, H., 2012. Detection of leakage freshwater and friction factor calibration in drinking networks using central force optimization. Water Resources Management 26 (8), 2347–2363.

Hansen, P., Mladenovic, N., 2001. Variable neighbourhood search: principles and applications. European Journal of Operational Research 130 (3), 449–467.

Hasançebi, O., Azad, S., 2013. Reformulations of Big Bang-Big Crunch algorithm for discrete structural design optimization. World Academy of Science, Engineering and Technology, International Science Index 74 International Journal of Civil, Environmental, Structural, Construction and Architectural Engineering 7 (2), 139–150.

Hassan, R., Cohanim, B., de Weck, O., Venter, G., 2005. A comparison of particle swarm optimization and the genetic algorithm. In: 46th AIAA/ASME/ASCE/AHS/ASC Structures, Structural Dynamics and Materials Conference. Austin, Texas, AIAA 2005-1897, pp. 1–13.

Hezer, S., Kara, Y., 2012. Solving vehicle routing problems with simultaneous delivery and pickup using bacterial foraging. In: Proceedings of the 41st International Conference on Computing and Industrial Engineering. Los Angeles, California, USA, pp. 380–385.

Holland, J., 1992. Adaptation in Natural and Artificial Systems, 2nd ed.. University of Michigan Press.

Hossain, M., El-shafie, A., 2013. Application of artificial bee colony (ABC) algorithm in search of optimal release of Aswan High Dam. Journal of Physics: Conference Series 423, 012001.

Hu, X., Eberhart, R., 2002. Solving constrained nonlinear optimization problems with particle swarm optimization. In: Proceedings of the Sixth World Multiconference on Systemics, Cybernetics and Informatics, vol. 5. Orlando, USA, pp. 203–206.

Jalali, M., Afshar, A., Marino, M., 2005. Ant colony optimization (ACO): a new heuristic approach for engineering optimization. In: Proceedings of the 6th WSEAS International Conference on Evolutionary Computing. Lisbon, Portugal, pp. 188–192.

Jamil, M., Yang, X.S., 2013. A literature search of benchmark functions for global optimization problems. International Journal of Mathematical Modelling and Numerical Optimization 4 (2), 150–194.

Kang, F., Li, J.J., Xu, Q., 2012. Damage detection based on improved particle swarm optimization using vibration data. Applied Soft Computing 12 (8), 2329–2335.

Karaboga, D., 2005. An Idea Based on Honey Bee Swarm for Numerical Optimization. Technical Report TR06. Computer Engineering Department, Engineering Faculty, Ericiyes University.

Karaboga, D., Basturk, B., 2007. A powerful and efficient algorithm for numerical function optimization: artificial bee colony (ABC) algorithm. Journal of Global Optimization 39 (3), 459–471.

Karmarkar, N., 1984. A new polynomial time algorithm for linear programming. Combinatorica 4 (4), 373–395.

Kaveh, A., Mahdavi, V., 2014. Colliding bodies optimization: a novel meta-heuristic method. Computers and Structures 139, 18–27.

Kaveh, A., Talatahari, S., 2009. Size optimization of space trusses using Big Bang-Big Crunch algorithm. Computers and Structures 87 (17–18), 1129–1140.

Kennedy, J., Eberhart, R., 1995. Particle swarm optimization. In: Proceedings of the IEEE International Conference on Neural Networks, vol. 4. Piscataway, NJ, pp. 1942–1948.

Kiehbadroudinezhad, S., Noordin, N., Sali, A., Abidin, Z., 2014. Optimization of an antenna array using genetic algorithms. The Astronomical Journal 147 (6), 147.

Kirkpatrick, S., Gelett, C., Vecchi, M., 1983. Optimization by simulated annealing. Science 220 (4598), 671–680.

Kyprianou, A., Worden, K., Panet, M., 2001. Identification of hysteretic systems using the differential evolution algorithm. Journal of Sound and Vibration 248 (2), 289–314.

Leccardi, M., 2005. Comparison of three algorithms for Lévy noise generation. In: Proceedings of the 5th EUROMECH Non Linear Dynamics Conference.

Li, B., Gong, L., Yang, W., 2014. An improved artificial bee colony algorithm based on balance-evolution strategy for unmanned combat aerial vehicle path planning. The Scientific World Journal 2014, 232704.

Li, C., Zhou, J., 2011. Parameters identification of hydraulic turbine governing system using improved gravitational search algorithm. Energy Conversion and Management 52 (1), 374–381.

Liu, Y., Passino, K., 2002. Biomimicry of social foraging bacteria for distributed optimisation: models principles and emergent behaviours. Journal of Optimization Theory and Applications 115 (3), 603–628.

Liu, H., Su, R., Gao, Y., Xu, R., 2009. Improved PSO algorithm using two novel parallel inertia weights. In: Proceedings of Second IEEE Conference on Intelligent Computing Technology and Automation. Hunan, China, pp. 185–188.

Liu, Y., Tian, P., 2015. A multi-start central force optimization for global optimization. Applied Soft Computing 27, 92–98.

Mangaraj, B., Misra, I., Sanyal, S., 2011. Application of bacteria foraging algorithm for the design optimization of multi-objective Yagi-Uda array. International Journal of RF and Microwave Computer-Aided Engineering 21 (1), 25–35.

Mangaraj, B., Misra, I., Sanyal, S., 2013. Application on bacterial foraging algorithm in designing log periodic dipole array for entire UHF TV spectrum. International Journal of RF and Microwave Computer-Aided Engineering 23 (2), 157–171.

Mantegna, R., 1994. Fast accurate algorithm for numerical simulation of Lévy stable stochastic processes. Physical Review E 49 (5), 4677–4683.

Markowitz, H., 1952. Portfolio selection. The Journal of Finance 7 (1), 77–91.

Meruane, V., Heylen, W., 2011. An hybrid real genetic algorithm to detect structural damage using model properties. Mechanical Systems and Signal Processing 25 (5), 1559–1573.

Michalewicz, Z., 1995. A survey of constraint handling techniques in evolutionary computation methods. In: Proceedings of the 4th Annual Conference on Evolutionary Programming.

Mirzaali, M., Seyedkashi, S., Liaghat, G., Naeini, H., Shojaee G, K., Moon, Y., 2012. Application of simulated annealing method to pressure and force loading optimization in tube hydroforming process. International Journal of Mechanical Sciences 55 (1), 78–84.

Mohamad, A., Zain, A., Bazin, N., 2014. Cuckoo search algorithm for optimization problems—a literature review and its applications. Applied Artificial Intelligence 28 (5), 419–448.

Napoles, G., Gran, I., Bello, R., 2012. Constricted particle swarm optimization algorithms for global optimization. Polibits 46, 5–11.

Narendar, S., Annadha, T., 2012. A hybrid foraging algorithm for solving job scheduling problems. International Journal of Programming Languages and Applications 2 (4), 1–11.

Narzizi, G., 2008. Classic Methods for Multi-Objective Function Optimization. Courant Institute of Mathematical Science, New York University. 31 January, 2008.

Passino, K., 2002. Biomimicry of bacterial foraging for distributed optimization and control. IEEE Control Systems Magazine 22 (3), 56–67.

Peng, B., Liu, B., Zhang, F-Y., Wang, L., 2009. Differential evolution algorithm-based parameter estimation for chaotic systems. Chaos, Solitons and Fractals 39 (5), 2110–2118.

Pereyra, V., Saunders, M., Castillo, J., 2013. Equispaced Pareto front construction for constrained bi-objective optimization. Mathematical and Computer Modelling 57 (9–10), 2122–2131.

Plagianakos, V., Tasoulis, D., Vrahatis, M., 2008. A review of major application areas of differential evolution. In: Advances in Differential Evolution. In: Studies in Computational Intelligence, vol. 143. Springer, pp. 197–238.

Quaraab, A., Ahiod, B., Yang, X.S., 2014. Discrete cuckoo search algorithm for the travelling salesman problem. Neural Computing and Applications 24 (7–8), 1659–1669.

Qubati, G., Formato, R., Dib, N., 2010. Antenna benchmark performance and array synthesis using central force optimization IET microwaves. Antennas and Propagation 4 (5), 583–592.

Rahnamayan, S., Tizhoosh, H., Salama, M., 2008. Opposition-based differential evolution. IEEE Transactions on Evolutionary Computation 12 (1), 64–79.

Rajeev, S., Krishnamoorthy, C., 1992. Discrete optimization of structures using genetic algorithms. Journal of Structural Engineering 118 (5), 1233–1250.

Rajpaul, V., 2012. Genetic algorithms in astronomy and astrophysics. In: Proceedings the 56th Annual Conference of the South African Institute of Physics, pp. 519–524.

Rashedi, E., Nezamabadi-Pour, H., Saryazdi, S., 2009. GSA: a gravitational search algorithm. Information Sciences 179 (13), 2232–2248.

Rashedi, E., Nezamabadi-Pour, H., Saryazdi, S., 2011. Filter modeling using gravitational search algorithm. Engineering Applications of Artificial Intelligence 24 (1), 117–122.

Ratnaweera, A., Halgamuge, S., Watson, H., 2004. Particle swarm optimizer with time-varying acceleration coefficients. IEEE Transactions on Evolutionary Computation 8 (3), 240–255.

Reddy, D., Chandra-Sekhar, J., 2014. Application of firefly algorithm for combined economic load and emission dispatch. International Journal on Recent and Innovation Trends in Computing and Communication 2 (8), 2448–2452.

Salhi, A., Fraga, E., 2011. Nature inspired optimization approaches and the new plant propagation algorithm. In: Proceeding of the International Conference on Numerical Analysis and Optimization. ICeMATH'11, Yogyakarta, Indonesia, pp. K2-1–K2-8.

Schutte, J., Groenwold, A., 2003. Sizing design of truss structures using particle swarms. Structural and Multidisciplinary Optimization 25 (4), 261–269.

Shah-Hosseini, H., 2008. Optimization with the nature-inspired intelligent water drops algorithm. International Journal of Intelligent Computing and Cybernetics 1 (2), 193–212.

Shahzad, F., Baig, A., Masood, S., Kamran, M., Naveed, N., 2009. Opposition—based particle swarm optimization with velocity clamping (OVCPSO). In: Advances in Computational Intelligence, pp. 339–348.

Shi, Y., Eberhart, R., 1998. A modified particle swarm optimizer. In: Proceedings of the IEEE Congress on Evolutionary Computation, pp. 69–73.

Siddique, N., Adeli, H., 2015. Nature inspired computing: an overview and some future directions. Cognative Computation 7 (6), 706–714.

Storn, R., Price, K., 1997. Differential evolution—a simple and efficient heuristic for global optimization over continuous spaces. Journal of Global Optimization 11 (4), 341–359.

Stuzle, T., Hoos, H., 1997. Max-min ant system and local search for the travelling salesman problem. In: IEEE International Conference on Evolutionary Computation. ICEC'97, Piscataway, NJ, USA, pp. 309–314.

Surjanovic, S., Bingham, D., 2013. Virtual library of simulation experiments: test functions and datasets. Retrieved January 23, 2017, from http://www.sfu.ca/~ssurjano.

Tabrizian, Z., Afshari, E., Amiri, G., Ali Beigy, M., Nejad, S., 2013. A new damage detection method: Big Bang-Big Crunch (BB-BC) algorithm. Shock and Vibration 20 (4), 633–648.

Taherian, H., Nazer, I., Razavi, E., Goldani, S., Farshad, M., Aghaebrahimi, M., 2013. Application of an improved neural network using cuckoo search algorithm in short-term electricity price forecasting under competitive power markets. The Journal of Operation and Automation in Power Engineering 1 (2), 136–146.

Talukder, S., 2011. Mathematical Modelling and Applications of Particle Swarm Optimization. MSc Thesis, SE-371 79. School of Engineering, Blekinge Institute of Technology, Karlskrona, Sweden.

Tawfik, A., Badr, A., Abdel-Rahman, I., 2013. One rank cuckoo search algorithm with application to algorithmic trading systems optimization. International Journal of Computer Applications 64 (6), 30–37.

Uppal, R., Kumar, S., 2016. Big Bang-Big Crunch algorithm for dynamic deployment of wireless sensor network. International Journal of Electrical and Computer Engineering 6 (2), 596–601.

Valian, E., Mohanna, S., Tavokoli, S., 2011. Improved cuckoo search algorithm for feedforward neural network training. International Journal of Artificial Intelligence and Applications 2 (3), 36–43.

Vasan, A., Simonovic, S., 2010. Optimization of water distribution network design using differential evolution. Journal of Water Resources Planning and Management 136 (2).

Weyland, D., 2010. A rigorous analysis of the harmony search algorithm: how the research community can be misled by a 'Novel' methodology. International Journal of Applied Metaheuristic Computing 1 (2), 50–60.

Wolpert, D., Macready, W., 1997. No free lunch theorems for optimization. IEEE Transactions on Evolutionary Computation 1 (1), 67–82.

Xin, J., Chen, G., Hai, Y., 2009. A particle swarm optimizer with multi-stage linearly decreasing inertial weight. In: Proceedings of the International Joint Conference on Computational Science and Optimization. Sanya, China, pp. 506–508.

Yaduwanshi, S., Sidhu, J., 2013. Application of bacterial foraging optimization as a de-noising filter. International Journal of Engineering Trends and Technology 4 (7), 3049–3055.

Yang, X.S., 2009. Firefly algorithms for multi-modal optimisation. In: Watanabe, O., Zeugmann, T. (Eds.), Proc. 5th Symposium on Stochastic Algorithms, Foundations and Applications. In: Lecture Notes in Computer Science, vol. 5792, pp. 169–178.

Yang, X.S., 2010. A new metaheuristic bat inspired algorithm. In: Nature Inspired Cooperative Strategies for Optimization. In: Studies in Computational Intelligence, vol. 284, pp. 65–73.

Yang, X.S., Deb, S., 2009. Cuckoo search via Lévy flights. In: Proceedings Sixth of World Congress on Nature and Biologically Inspired Computing. IEEE Publications, USA, pp. 210–214.

Yang, X.S., Deb, S., 2014. Cuckoo search: recent advances and applications. Neural Computing and Applications 24 (1), 169–174.

Yang, X.S., He, X., 2013. Firefly algorithm: recent advances and applications. International Journal of Swarm Intelligence 1 (1), 36–50.

Yeniay, O., 2005. Penalty function methods for constrained optimization with genetic algorithms. Mathematics and Computational Applications 10 (1), 45–56.

Yesil, E., Urbas, L., 2010. Big Bang-Big Crunch learning method for fuzzy cognitive maps. World Academy of Science, Engineering and Technology, International Science Index 47 International Journal of Computer, Electrical, Automation, Control and Information Engineering 4 (11), 1756–1765.

Yin, M., Hu, Y., Yang, F., Li, X., Gu, W., 2011. A novel hybrid K-harmonic means and gravitational search algorithm approach for clustering. Expert Systems with Applications 38 (8), 9319–9324.

Yu, Q., You, X., Liu, S., 2014. An improved ant colony algorithm based on 3-OPT and chaos for travelling salesman problem. International Journal on Cybernetics and Informatics 3 (5), 1–10.

INDEX

Printed in the United States
By Bookmasters